Code of safe working practices for merchant seamen

London: HMSO

©Crown copyright 1991
First published 1991

ISBN 0 11 551048 6

Preface

This Code of Safe Working Practices is intended primarily for merchant seamen on United Kingdom registered vessels.

Copies of the current edition of the Code must be carried on all United Kingdom ships other than fishing vessels and pleasure craft, and a copy must be made available to any seaman in the ship who requests it, in accordance with Regulation 3 of the Merchant Shipping (Code of Safe Working Practices) Regulations 1980 (SI 1980 No 686). If no more than 15 persons (Master and seamen) are employed on a ship, the minimum number of copies that the ship may carry is two, of which one should be kept in the custody of the Master and one in a place readily accessible to seamen. Where the complement of seamen is more than 15, further copies must be carried in accordance with the Regulation.

This edition incorporates amendments issued between August 1988 and November 1990 in a series of 7 Merchant Shipping Notices, and supersedes all previous editions.

Marine Directorate,
Department of Transport,
London.
August 1991

Contents

1	**GENERAL**	1
1.1	Introduction	1
1.2	Health and hygiene	3
1.3	Working clothes	4
1.4	Shipboard housekeeping	5
1.5	Substances hazardous to health	6
1.6	Asbestos dust	7
2	**FIRE PRECAUTIONS**	9
2.1	Smoking	9
2.2	Electrical and other fittings	9
2.3	Laundry and wet clothing	10
2.4	Spontaneous combustion	10
2.5	Machinery spaces	11
2.6	Galleys	11
2.7	Hot work	11
3	**EMERGENCY PROCEDURES**	12
3.1	Musters and drills	12
3.2	Fire drills	13
3.3	Survival craft drills	13
3.4	Action in the event of fire	15
4	**SAFETY OFFICIALS**	17
4.1	Introduction and Application	17
4.2	Appointment of Safety Officers	18
4.3	Election of safety representatives	19
4.4	Safety committees	19
4.5	Termination of appointments	21
4.6	Advice to Safety Officer	21
4.7	Advice to safety representatives	25
4.8	Advice to safety committees	26
4.9	Advice to the employer and Master	26
4.10	Accident investigation	27
5	**PROTECTIVE CLOTHING AND EQUIPMENT**	31
5.1	General	31
5.2	Head protection	32
5.3	Hearing protection	32
5.4	Face and eye protection	33
5.5	Respiratory protective equipment	33
5.6	Hand and foot protection	34
5.7	Protection from falls	35
5.8	Body protection	35
5.9	Protection against drowning	35

6	SIGNS, NOTICES AND COLOUR CODES	36
6.1	General	36
6.2	Signs and notices	36
6.3	Portable fire extinguishers	37
6.4	Electrical wiring	37
6.5	Gas cylinders	37
6.6	Pipelines	38
6.7	Dangerous goods	39
7	**PERMIT TO WORK SYSTEMS**	**40**
8	**MEANS OF ACCESS**	**43**
8.1	General	43
8.2	Standards of construction	43
8.3	Maintenance	44
8.4	Positioning of access equipment	44
8.5	Lighting and safety of movement	45
8.6	Portable and rope ladders	45
8.7	Safety nets	45
8.8	Life-buoys	46
8.9	Special circumstances and general guidance	46
8.10	Corrosion of accommodation ladders and gangways	46
9	**SAFE MOVEMENT ON BOARD SHIP**	**48**
9.1	General	48
9.2	Transit areas	48
9.3	Lighting	49
9.4	Safety signs	50
9.5	Guarding of openings	50
9.6	Ladders	51
9.7	Vehicles	52
9.8	Drainage	53
9.9	Watertight doors	53
9.10	General advice to seafarers	53
10	**ENTERING ENCLOSED OR CONFINED SPACES**	**54**
10.1	General	54
10.2	Precautions on entering dangerous spaces	54
10.3	Duties and responsibilities of a competent person and a responsible officer	55
10.4	Identifying potential hazards	55
10.5	Preparing and securing the space for entry	57
10.6	Testing the atmosphere of the space	58
10.7	Use of a permit-to-work system	59
10.8	Procedures and arrangements before entry	59
10.9	Procedures and arrangements during entry	60
10.10	Additional requirements for entry into a space where the atmosphere is suspect or known to be unsafe	61
10.11	Drills and rescue	61
10.12	Breathing apparatus and resuscitation equipment	62
10.13	Maintenance of equipment	64
10.14	Training, instruction and information	64
10.15	Statutory regulations	64
11	**MANUAL LIFTING AND CARRYING**	**65**
11.1	Guidance to employers	65
11.2	Guidance to seafarers	65

12	TOOLS AND MATERIALS	67
12.1	Work equipment and the employer	67
12.2	Hand tools	67
12.3	Portable electric, pneumatic and hydraulic tools and appliances	68
12.4	Workshop and bench machines (fixed installations)	69
12.5	Abrasive wheels	70
12.6	Spirit lamps	71
12.7	Compressed air	71
12.8	Compressed gas cylinders	72
12.9	Chemical agents	72
13	WELDING AND FLAMECUTTING OPERATIONS	73
13.1	General	73
13.2	Protective clothing	73
13.3	Precautions against fire and explosion	74
13.4	Electric welding equipment	74
13.5	Precautions to be taken during electric-arc welding	75
13.6	Gas welding and cutting	76
14	PAINTING	77
14.1	General	77
14.2	Spraying	77
14.3	Painting aloft, overside and from punts	79
15	WORKING ALOFT AND OUTBOARD	80
15.1	General	80
15.2	Cradles and stages	81
15.3	Bosun's chair	82
15.4	Ropes	82
15.5	Portable ladders	84
16	ANCHORING, CASTING OFF AND TOWING	85
16.1	Anchoring	85
16.2	Characteristics of man-made fibre ropes	85
16.3	Making fast	87
16.4	Mooring to buoys	88
16.5	Casting off	88
16.6	Towing	88
17	LIFTING PLANT	90
17.1	Introduction	90
17.2	Use of lifting plant	90
17.3	Use of winches and cranes	94
17.4	Winches	94
17.5	Cranes	95
17.6	Use of derricks	95
17.7	Use of derricks in union purchase	96
17.8	Use of stoppers	97
17.9	Overhaul of cargo gear	98
17.10	Trucks and other mechanical handling appliances	98
17.11	Defects	98
17.12	Testing of lifting plant	99
17.13	Examination of lifting plant	100
17.14	Marking of lifting appliances and gear	100
17.15	Certificates and reports	101
18	HATCHES	102
18.1	Introduction	102
18.2	General	102

18.3	Mechanical hatch covers	103
18.4	Non-mechanical hatch covers and beams	103
18.5	Steel-hinged inspection/access lids	104
19	**WORK IN CARGO SPACES**	**106**
19.1	Access	106
19.2	Lighting in cargo spaces	106
19.3	Fencing	106
19.4	General precautions	107
20	**WORK IN MACHINERY SPACES**	**108**
20.1	General	108
20.2	Boilers	109
20.3	Unmanned machinery spaces	109
20.4	Refrigeration machinery	110
21	**HYDRAULIC AND PNEUMATIC EQUIPMENT**	**111**
21.1	General	111
21.2	Hydraulic jacks	112
22	**OVERHAUL OF MACHINERY**	**113**
22.1	General	113
22.2	Protective clothing and equipment	114
22.3	Lifting	115
22.4	Floor plates and handrails	115
22.5	Working aloft or over bottom platforms	115
22.6	Boilers	115
22.7	Auxiliary machinery and equipment	116
22.8	Main engines	116
22.9	Electrical equipment	117
22.10	Refrigeration machinery and refrigerated compartments	118
22.11	Steering gear	118
22.12	Hydraulic and pneumatic equipment	119
23	**SERVICING RADIO AND ASSOCIATED ELECTRONIC EQUIPMENT**	**120**
23.1	General	120
23.2	Electrical hazards	121
23.3	Valves and semi-conductor devices	121
23.4	Work on apparatus on extension runners or on the bench	122
23.5	Work with visual display units	122
24	**STORAGE BATTERIES**	**124**
24.1	General	124
24.2	Lead-acid batteries	125
24.3	Alkaline batteries	126
25	**WORK IN GALLEY, PANTRY AND OTHER FOOD HANDLING AREAS**	**127**
25.1	Health and hygiene	127
25.2	Slips, falls and tripping hazards	128
25.3	Galley stoves, steamboilers and deep fat fryers	128
25.4	Catering equipment	130
25.5	Knives, saws, choppers etc	130
25.6	Refrigerated rooms and store rooms	134
26	**WORK IN SHIPS' LAUNDRIES**	**133**
26.1	General	133
26.2	Burns and scalds	133
26.3	Machinery and equipment	133

26.4	Dry-cleaning operations	134
26.5	Fire prevention	135
27	**GENERAL CARGO SHIPS**	**136**
27.1	Stowage of cargo	136
27.2	Dangerous goods and substances	136
27.3	Working cargo	138
28	**CARRIAGE OF CONTAINERS**	**140**
29	**TANKERS AND OTHER SHIPS CARRYING BULK LIQUID CARGOES**	**141**
29.1	General	141
29.2	Oil and bulk ore/oil carriers	141
29.3	Liquefied gas carriers	142
29.4	Chemical carriers	143
30	**SHIPS SERVING OFFSHORE GAS AND OIL INSTALLATIONS**	**145**
30.1	General	145
30.2	Carriage of cargo on deck	146
30.3	Lifting, hauling and towing gear	146
30.4	Approaching rig and cargo handling at rig	146
30.5	Transfer of personnel from ship to rig by 'personnel baskets'	147
30.6	Transfer of personnel by boat	148
30.7	Anchor handling	148
APPENDIX 1	**LIST OF BRITISH STANDARD SPECIFICATIONS RELEVANT TO RECOMMENDED SAFE WORKING PRACTICES**	**149**
	1A Arranged by Code chapter	149
	1B Arranged by BS number	151
APPENDIX 2	**BIBLIOGRAPHY**	**153**
APPENDIX 3	**MERCHANT SHIPPING HEALTH AND SAFETY LAW**—A brief guide for seafarers	**156**
APPENDIX 4	**LIFTING PLANT**	
	4A Examples of hand signals for use with lifting appliances on ships	158
	4B Certificates of test and thorough examination	159
	4C Register of ships' lifting appliances and cargo handling gear	160
INDEX		**167**

CHAPTER 1

General

1.1 Introduction

1.1.1 This Code provides information and guidance on procedures to be followed and measures to be adopted for improving the safety and health of those living and working on board ship.

1.1.2 The Code is addressed to everyone on a ship regardless of rank or rating because the recommendations can be effective only if they are understood by all and if all co-operate in their implementation. Those not themselves actually engaged in a job in hand should be aware of what is being done, so that they may avoid putting themselves at risk or those concerned at risk by impeding or needlessly interfering with the conduct of the work.

1.1.3 This Code is arranged so that matters which concern every seafarer on every ship are dealt with in the earlier Chapters, then follows advice on different sorts of jobs common to most ships and lastly come provisions related to work particular to special kinds of ship.

1.1.4 The aim has been to state what is accepted as normal, safe practice but ships differ in construction, layout and equipment, in function and in conditions of service. In the circumstances of an individual ship, it may be impracticable or inadvisable for some reason to comply exactly with a recommended practice. In such instances, the Code provisions will usually indicate the nature of the principal risks and care should be taken to adopt an alternative method which adequately guards against those risks and which does not itself introduce any special new hazard.

1.1.5 The guidance covers safe working practices for most of the situations that commonly arise on ships. The experienced seafarer who uses the Code should be able to follow this guidance and also to adapt it and apply it in principle to many other work situations that the Code does not specifically cover.

1.1.6 Recommendations are necessarily framed in broad terms covering only aspects related to the particular operation described. It is the responsibility of those concerned in carrying out the operation to ensure that it is done in all other respects efficiently and safely even though details of every necessary precaution may not have been explicitly stated in the relevant provision of the Code.

1.1.7 Many of the provisions of this Code relate to matters which are the subject of UK statutory regulations. Such provisions are intended to give guidance on how the statutory obligations should be fulfilled. Certain of the regulations specifically require full account to be taken of the principles and guidance of the corresponding chapters of the Code (see paragraphs

8.1.1, 9.1.1, 10.15.1, 17.1.1 and 18.1.1). In no case, however, is the guidance to be regarded as supplanting or amending the relevant legal requirements.

MS (Code of Safe Working Practices) Regs SI 1980 No. 686

1.1.8 Owners of UK ships are required by Merchant Shipping Regulations to ensure that copies of this Code are carried on board ship. The minimum number of copies is specified according to the number of persons employed on a vessel. Masters should ensure that the Code is readily accessible to all crew members. It may be supplemented by other Codes, guides or instructions on safety that may be issued by companies to particular ships.

MS (Health and Safety: General Duties) Regs SI 1984 No. 408

1.1.9 Merchant Shipping Regulations place general duties on employers, employees and others in respect of health and safety on board UK ships. The main duties are as follows:
(a) on employers—to ensure, so far as is reasonably practicable, the health and safety of their employees and other persons aboard ship;
(b) on employees—to take reasonable care for their own health and safety and for others aboard ship and to co-operate with their employers with regard to their statutory health and safety duties.

These regulations also make it an offence for any person to intentionally or recklessly interfere with or misuse anything provided in the interests of health and safety aboard a UK ship in pursuance of Merchant Shipping legislation.

1.1.10 Further information for seafarers and their employers on United Kingdom maritime safety legislation is contained in Appendix 3 and has also been issued as a Merchant Shipping Notice.

MSN No. M1398

1.1.11 Owners, managers, employers and shore staff should ensure that those on board the ships for which they are responsible are not subjected to unnecessary risk to health or safety. The employer or the Master should ensure that every risk to the health or safety of persons working on board ship is properly evaluated. Where they are able to remove a health or safety hazard at source they should do so. Failing this they should ensure that adequate measures (including the provision of any necessary safety equipment, and of training and instruction) are taken so as to reduce the risk to the lowest practicable level.

1.1.12 Although the importance of evaluating risks, removing or controlling hazards and providing suitable health and safety training is repeated elsewhere in this Code in respect of specific risks and hazards the principle also has a far broader application to all areas of occupational health and safety. Failure to protect workers by paying sufficient attention to a risk to health or safety could result in prosecution.

1.1.13 Seafarers should care for their own health and safety and for the health and safety of others. Any serious or imminent danger should be reported immediately to the appropriate officer. Proper use should be made of plant and machinery, and any hazard to health and safety (such as a dangerous substance) treated with due caution. When appropriate, seafarers should use the protective clothing and equipment provided for them, and afterwards return it to its proper place. No seafarer should do anything that would place any person at risk. For example, no one should disconnect, remove or interfere with any safety device without proper authority.

1.1.14 Accidents have often resulted from seemingly minor causes and the seriousness of the consequences is often a matter of chance. Any incident, though apparently trivial, should be regarded as a warning of something wrong in the system of work, the equipment used or the working area.

Immediate attention should be given to remedying the defect or deficiency in order to avoid a repetition of the incident which might have more serious consequences.

1.1.15 Where a ship or an item of her equipment has novel features, consideration should be given to any hazards these present as they may require special precautions to be taken. Any information or working instructions provided should be carefully studied to that end.

1.2 Health and hygiene

1.2.1 It is the seafarer's responsibility to look after his own health and fitness. High standards of personal cleanliness and hygiene should be maintained.

1.2.2 Good health depends on an even and thoughtful balance of work, rest and active play, on sensible and regular meals, on adequate sleep and an avoidance of excesses of rich food, alcohol and tobacco.

1.2.3 On board ship, simple infections can easily be spread from one person to others. Thus preventive measures, as well as easily effective treatment, are essential.

1.2.4 Cuts and abrasions should be cleansed at once and given first aid treatment as necessary to protect against infection.

1.2.5 Many serious infections can be guarded against by inoculation and vaccination. These should be kept up to date as necessary to meet the requirements of the international voyages to be undertaken.

MSN No. M1319 Sch 3

1.2.6 The risk of contracting malaria in certain parts of the world can be much reduced by taking precautions to avoid bites from mosquitoes carrying the disease, eg by the use of mosquito wire-screening and nets, keeping openings closed, and the use of anti-mosquito preparations or insecticides. Clothing also affords a degree of protection against mosquito bites and seafarers should therefore avoid going about after dusk with any part of the arms or legs exposed. On a ship bound for a malarious port all members of the crew should also take appropriate anti-malarial medication to control the risk of infection. The most effective treatment varies geographically according to the nature and resistance of local malarial germs. Further guidance is included in the relevant Merchant Shipping Notice.

1.2.7 Barrier creams may help to protect exposed skin against dermatitis and also make thorough cleaning easier.

1.2.8 Prolonged exposure to mineral oils may cause dermatitis and skin cancer. All traces of oil should be thoroughly washed from the skin but hydrocarbon solvents should be avoided. Working clothes should be laundered frequently. Oil-soaked rags should not be put in pockets.

1.2.9 Anthrax is a dangerous disease. It can be contracted by handling the hides, wool, bristles, bones, horns, hooves or other products from infected animals and from any wrapping materials which have contained them. Overalls, head covering and protective gloves should be worn to protect the skin as far as possible. Fuller information is given in Guidance Note EH 23 (Anthrax: health hazards) published by the Health and Safety Executive (See Appendix 2).

1.2.10   Rats and other rodents may be carriers of infection and should never be handled, dead or alive, with bare hands.

1.2.11   Inadvertent exposure to or contact with toxic chemicals or other harmful substances should be reported immediately and the appropriate remedial action taken.

1.2.12   Prolonged exposure to synthetic domestic cleaners and detergents is a potential cause of alkali (de-fatting) dermatitis. Cotton-lined rubber or PVC gloves should be worn when using such substances.

1.2.13   Some domestic substances, for example caustic soda and bleaching powders or liquids, can burn the skin. They may react dangerously with other substances and ought not to be mixed indiscriminately.

1.2.14   High humidity and heat can lead to heat exhaustion and heat stroke which may be fatal. When working in these conditions it is advisable to drink at least 4.5 litres (8 pints) of cool (but not iced) water daily. It is best to take small quantities at frequent intervals. Extra salt is essential; this can be in the form of two salt tablets four times a day or a level teaspoonful of table salt in plenty of water each morning and again in the evening, or added to food. If the work is in enclosed spaces, they should be well ventilated.

1.2.15   In tropical areas especially, exposure to the sun during the hottest part of the day should be avoided as far as possible. When it is necessary to work in very strong sunlight, appropriate clothing offering protection to both head and body should be worn, whatever the degree of acclimatisation may be.

1.2.16   Where it is required to work in exceptionally hot and/or humid conditions or when wearing respiratory equipment, it should be recognised that breaks at intervals in the fresh air or in the shade may be necessary.

1.2.17   Mis-use of alcohol or drugs affects a person's fitness for duty and harms his health. The immediate after-effects may increase liability to accidents. Drinking alcohol whilst under treatment with prescribed drugs should be avoided, since even common remedies such as aspirin, seasickness tablets or codeine may be dangerous in conjunction with alcohol.

1.2.18   As a general rule fresh fruit and salad should be thoroughly washed in fresh water before being eaten.

1.2.19   Coughs and lung damage can be caused by breathing irritant dust. This may be produced by many different substances to which the general guidance in Section 1.5 applies. Dust containing asbestos fibres is particularly hazardous since this can cause lung cancer and other serious lung diseases when inhaled (see Section 1.6 of this Code and the relevant Merchant Shipping Notice). The risk is usually much greater for a person who smokes than for a non-smoker.

MSN No. M1428

1.3   Working clothes

1.3.1   Clothing should be chosen to minimise working risks.

1.3.2   Working clothes should be close-fitting with no loose flaps, bulging pockets or ties, since injuries may result from clothing being caught up by moving parts of machinery or garments catching on obstructions or projections and causing falls. Clothing worn in galleys etc where there is a risk of burning or scalding should adequately cover the body to minimise this risk and be of a material of low flammability such as cotton or a cotton/terylene mix. Clothes should be kept in good repair.

1.3.3   Shirts or overalls provide better protection if they have long sleeves. Long sleeves should not be rolled up.

1.3.4   Scarves, sweat rags and other neck wear, loose clothing, finger rings and jewellery can be hazards when working with machinery. Long hair should be covered.

1.3.5   Sandals and plimsolls are dangerous and should not be worn when working, since they offer little protection against accidental scalds or burns or falling objects and add to the risks of tripping and falling or slipping on ladders (as do old, worn out, down-at-heel shoes). The wearing of appropriate industrial or safety footwear, which can be of good appearance, is recommended (see Section 5.6).

1.3.6   Gloves are a sensible precaution when handling sharp or hot objects but may easily be trapped on drum ends and in machinery. Whilst loose-fitting gloves allow hands to slip out readily, they do not give a good grip on ladders. Wet or oily gloves may be slippery and great care should be taken when working in them.

1.4   Shipboard housekeeping

1.4.1   Most of this Code deals with foreseeable risks arising in particular places and in performing particular jobs, but accidents may happen at any time in any part of the ship. Many such accidents can be prevented by always keeping things ship-shape and doing things in an orderly fashion.

1.4.2   Wear and tear on a ship in service give rise to minor deficiencies in the structure, equipment or furnishings; for example, protruding nails and screws, loose fittings and handles, uneven and damaged flooring, rough and splintered edges to woodwork and jamming doors, any of which may cause cuts, bruises or trips and falls. They should be put right as soon as they are noticed.

1.4.3   If asbestos-containing panels, cladding or insulation work loose or are damaged in the course of a voyage, pending proper repair, the exposed edges or surfaces should be protected by a suitable coating or covering to prevent asbestos fibres being released and dispersed in the air (see also section 1.6).

1.4.4   Flickering lights usually indicate faults in wiring or fittings which may cause electric shock or fires. They should be investigated and repaired by a competent person. Failed light bulbs should be replaced as soon as possible.

1.4.5   Instruction plates, notices and operating indicators should be kept clean and legible.

1.4.6   Heavy objects, particularly if at a height above deck level should be stowed securely against the movement of the ship or inadvertent displacement. Similarly, furniture etc likely to fall or shift during heavy weather should be properly secured.

1.4.7 Doors whether open or closed, should be properly secured; they should not be left swinging.

1.4.8 Litter may present a fire risk or cause slips or falls, but in any case may conceal some other hazard (see Paragraph 2.4.1).

1.4.9 Tidiness not only makes hidden defects apparent but ensures that articles are in their proper place to be found as required.

1.4.10 In carrying out any task, possible risks to other persons should be considered; for example, if water from a careless hosing-down of the deck enters a galley through an open light or scuttle, it may be most dangerous to galley staff.

1.4.11 Care is needed in personal matters. Dangerous articles such as razor blades and lighted cigarette ends should be disposed of safely.

1.4.12 Many aerosols have volatile and inflammable contents. They should never be used or placed near naked flames or other heat source even when 'empty'. Empty canisters should be properly disposed of.

1.4.13 Some fumigating or insecticidal sprays contain ingredients which though perhaps in themselves harmless to human beings, may be decomposed when heated. Smoking may be dangerous in sprayed atmospheres while the spray persists.

1.5 Substances hazardous to health

1.5.1 Many substances found on ships are capable of damaging the health of those exposed to them. They include not only substances displaying hazard warning labels (eg on dangerous goods cargoes and ships' stores) but also, for example, a range of dusts, fumes and fungal spores from goods, plant or activities aboard ship.

1.5.2 Whenever crew members work in the presence of substances hazardous to health the employer or the Master should ensure that any risks from exposure are assessed (see 1.5.4) and appropriate measures taken to prevent or control them (see 1.5.8–9). The assessment should include consideration of any necessary precautionary measures both for crew members and other affected groups (eg stevedores and maintenance personnel). Failure to protect workers exposed to hazardous substances in this way could result in prosecution.

1.5.3 Employers should instruct, inform and train crew so that they know and understand the risks arising from their work, the precautions to be taken and the results of any monitoring of exposure.

1.5.4 The health risk from a substance hazardous to health should be assessed by a competent person who should look at each hazard from each of the following viewpoints: the identity, concentration, form and possible harmful effects of the hazardous substance, including any harmful products; the likely exposure for the work in hand; and the number of people (crew members and others) who will be in contact (they should be identified). The risks to crew members and other persons should be considered separately for each area of the ship and each risk should be assessed individually. Where appropriate the risks to different categories of person should be considered separately.

1.5.5 A risk will normally conform to one or other of the following categories:—
(a) insignificant—no further action needed unless conditions change;
(b) significant—immediate and long-term control measures needed;
(c) at present under control—precautions needed to maintain control or to regain it if risk should later increase; or
(d) uncertain or unknown—degree of exposure or risk from exposure needs to be established, if necessary with the help of outside experts; meanwhile caution is required.

1.5.6 As an aid to assessment of the risks from dangerous goods reference may be made to the International Maritime Dangerous Goods Code (see 27.2.1) or to the Chemical data sheets contained in the Tanker Safety Guides (Gas and Chemical) issued by the International Chamber of Shipping. Information concerning hazardous cargoes carried in bulk should be available where applicable to allow the assessment to be made. In the case of ship's stores etc, reference should be made to manufacturer's instructions and data sheets. Reference may also be made where appropriate to the series of publications issued by the Health and Safety Executive under the Control of Substances Hazardous to Health Regulations (see Appendix 2).

Control of Substances Hazardous to Health Regs SI 1988 No 1657 ('COSHH' Regs)

1.5.7 The assessment should be reviewed regularly and revised if there is a significant change of circumstances. A record should normally be kept of the assessment and of any measures taken. It is recommended that such records should be stored together on board to form a single database.

1.5.8 Prevention of the hazard by not having the substance is always better than control; but failing this the control measures should achieve adequate control, be used, and be maintained in efficient order. When adequate control is not feasible by any other means, then as a means of last resort personal protective clothing and equipment should be provided and used instead: see Chaper 5.

1.5.9 For certain substances (eg where the risk to health is through inhalation) very specific control measures apply: for example, where the substance is asbestos dust (see 1.6) or a dangerous gas (see 27.2.10), or is in an enclosed or confined space (see Chapter 10). Where the risk is of a lower order, effective controls will often be the simplest however, eg awareness of the problem and an organised working method to reduce exposure.

1.5.10 In cases where failure of the control measures could result in serious risks to health, or where their adequacy or efficiency is in doubt, the exposure of crew members should be monitored and a record kept for future reference.

1.6 Asbestos Dust

1.6.1 All types of asbestos have a fibrous structure and can produce harmful dust if the surface exposed to the air is damaged or disturbed. The danger is not immediately obvious because the fibres which damage the lungs and can cause lung cancer are too small to be seen with the naked eye. Asbestos which is in good condition is unlikely to release fibres, but where the material is damaged or deteriorating, or work is undertaken on it, airborne fibres can be released. Dry asbestos is much more likely to produce dust than asbestos that is thoroughly wet or oil-soaked. Asbestos is particularly likely to occur on older vessels in old insulation and panelling, but certain asbestos compounds may also be found on other UK vessels in machinery components such as gaskets and brake linings.

1.6.2   Shipowners should advise Masters of any location where asbestos is known or believed to be present on their ship. Masters and/or safety officers should keep a written record of this information and should also note any other position where asbestos is suspected, but they should not probe or disturb any suspect substance. Crew members who work regularly near asbestos or a substance likely to contain it should be warned of the need for caution and should report any deterioration in its condition such as cracking or flaking.

1.6.3   The condition of old asbestos may deteriorate and where reasonably practicable consideration should be given to its removal. This should be carried out in port and to ensure the use of adequate protective procedures a specialist asbestos removal contractor should be used. Where the port is in the UK and the work involves asbestos insulation or asbestos coating it is usually necessary for the contractor to hold a licence issued by the Health and Safety Executive. If such work is carried out outside the UK the contractor should be of equivalent competence.

MSN No. M1428

1.6.4   If it is essential to carry out emergency repairs liable to create asbestos dust while the ship is at sea strict precautions, including the use of the appropriate protective clothing and respiratory protective equipment, should be observed in accordance with the guidance given in the relevant Merchant Shipping Notice. See also the general guidance on the assessment and control of risks from hazardous substances in Section 1.5 of this Code.

MSN No. M1428 S.8

1.6.5   Guidance on precautions to be taken when asbestos is carried as a cargo is also included in a Merchant Shipping Notice.

CHAPTER 2

Fire Precautions

The only sure way to avoid the disastrous consequences of a fire at sea is not to have one. All on board have therefore a personal interest in observing all practicable precautions against the outbreak of fire.

2.1 Smoking

2.1.1 Fires are often caused by the careless disposal of burning cigarette ends and matches. Ashtrays or other suitable containers should be provided and used at places where smoking is authorised. Care should be taken to ensure that matches are actually extinguished and cigarette ends properly stubbed out. They should not be thrown overboard since there is a danger that they may be blown back on board.

2.1.2 Conspicuous warning notices should be displayed in any part of the ship where smoking is forbidden (permanently or temporarily) and these should be obeyed in all circumstances.

2.1.3 It is dangerous to smoke in bed.

2.2 Electrical and other fittings

2.2.1 Unauthorised persons should not interfere with electrical fittings. Personal electrical appliances should be connected to the ship's supply only with the approval of the electrical officer or the responsible engineer officer. Notices should be displayed on the notice boards to this effect.

2.2.2 Faulty appliances, fittings or wiring which are part of the ship's equipment should be reported immediately to the head of department.

2.2.3 All electrical appliances should be firmly secured and served by permanent connections whenever possible.

2.2.4 Flexible leads should be as short as practicable and so arranged as to prevent their being chafed or cut in service.

2.2.5 Makeshift plugs, sockets and fuses should not be used.

2.2.6 Circuits should not be overloaded since this causes the wires to overheat, destroying insulation and thus resulting in a possible short-circuit which could start a fire.

2.2.7 All portable electrical appliances, lights etc should be isolated from the mains after use.

2.2.8   It is important that all fixed electric heaters are fitted with suitable guards securely attached to the heater and that the guards are maintained in position at all times. Temporary arrangements to hang clothing above the heaters or to dry clothing on the heaters should not be permitted and drying of clothing should only be carried out by using suitably designed equipment.

2.2.9   When using drying cabinets or similar appliances care should be taken so that the ventilation apertures are not obscured by overfilling of the drying space. As the ventilation apertures of drying appliances may become blocked due to accumulations of fluff from clothing, any screens or fine mesh covers associated with the ventilation apertures should be regularly inspected and cleaned.

2.2.10   The use of portable heaters should be avoided. However, if they are used with the ship in port (as temporary heating during repairs and as additional heating during inclement weather), the heaters should not be positioned on wooden floors or bulkheads, carpets or linoleum without the provision of a protective sheet of a non-combustible material. Portable heaters shoud be provided with suitable guards and care should be exercised when positioning the heater in relation to furniture and other fittings in the cabin or other space. Again, drying arrangements in relation to these heaters should not be permitted.

2.2.11   Personal portable space-heating appliances of any sort should not be used at sea and notices to this effect should be displayed on notice boards.

2.2.12   The construction and installation of electric heaters in merchant ships and fishing vessels should take due account, as appropriate, of the requirements of the relevant Rules and Regulations as expanded by the various Instructions and Guidance Notes where appropriate.

2.2.13   Permanent electric heaters are normally supplied with installation instructions by the manufacturers and these should be carefully followed.

2.3   Laundry and wet clothing

2.3.1   Clothing or other articles should not be placed over space heaters, or so close to heaters or light bulbs etc as to restrict the flow of air, and thus lead to overheating and fire (see also Section 26.5).

2.4   Spontaneous combustion

2.4.1   Dirty waste, rags, sawdust and other rubbish—especially if contaminated with oil—are dangerous if left lying about. Heat may be generated spontaneously within such rubbish which may be sufficient to ignite flammable mixtures or may become hot enough to set the rubbish itself on fire. Such waste and rubbish should therefore be properly stored until it can be safely disposed of as soon as possible thereafter.

2.4.2   Materials in ships' stores, including linen, blankets and similar absorbent materials are also liable to ignite by spontaneous combustion if damp or contaminated by oil. Strict vigilance, careful stowage and suitable ventilation are necessary to guard against such a possibility. If such materials become damp, they should be dried before being stowed away. If oil has soaked into them, they should be cleaned and dried, or destroyed. They should not be stowed in close proximity to oil or paints, or on or near to steam pipes.

2.5 Machinery spaces

2.5.1   The seriousness of fire in machinery spaces cannot be overstressed. All personnel should be fully aware of the precautions necessary for its prevention. Such precautions should include the maintenance of clean conditions, the prevention of oil leakage and the removal of all combustible materials from vulnerable positions (see Chapters 20 and 22).

2.5.2   Suitable metal containers should be provided for the storage of cotton waste, cleaning rags or similar materials after use. Such containers should be emptied at frequent intervals and the contents safely disposed of.

2.5.3   Wood, paints, spirits and tins of oil should not be kept in boiler rooms or machinery spaces.

2.5.4   All electric wiring should be well maintained and kept clean and dry. The rated load capacity of the wires and fuses should never be exceeded.

2.6 Galleys

2.6.1   Galleys and pantries present particular fire risks (see Chapter 25). Care should be taken in particular to avoid overheating or spilling fat or oil and to ensure that burners or heating plates are shut off when cooking is finished. Extractor flues and ranges etc should always be kept clean.

2.6.2   Means to smother fat or cooking oil fires, such as a fire blanket, should be readily available close to stoves.

2.7 Hot work

2.7.1   The precautions set out in Section 13.3 should be strictly followed to avoid the possibility of fire during welding, flame cutting or other hot work.

CHAPTER 3

Emergency Procedures

3.1 Musters and drills

MS (Musters and Training) Regs SI 1986 No. 1071. MSN Nos M 1217 and M 1396.

3.1.1 Musters and drills are required to be carried out regularly in accordance with Merchant Shipping Regulations. The guidance contained in this Chapter should be read in conjunction with information and guidance on these regulations issued in the relevant Merchant Shipping Notices.

3.1.2 Musters and drills have the objective of preparing a trained and organised response to situations of great difficulty which may unexpectedly threaten loss of life at sea. It is important that they should be carried out realistically, approaching as closely as possible to emergency conditions. Changes in the ship's function and changes in the ship's personnel from time to time should be reflected in corresponding changes in the muster arrangements.

3.1.3 The muster list should be conspicuously posted before the ship sails and, on international voyages and in ships of Classes IIA and III should be supplemented by emergency instructions for each crew member (eg in the form of a card issued to each crew member or affixed to individual crew berths or bunks). These instructions should describe the allocated muster station, survival craft station and emergency duty and all emergency signals and action, if any, to be taken on hearing such signals.

3.1.4 An abandon ship drill and a fire drill must be held within 24 hours of leaving port if more than 25% of the crew have not taken part in drills on board the ship in the previous month. As soon as possible but not later than two weeks after joining the ship, onboard training in the use of the ship's life-saving appliances, including survival craft equipment, should be given to crew members. As soon as possible after joining the ship, crew members should also familiarise themselves with their emergency duties, the significance of the various alarm signals and the locations of their lifeboat station and of all lifesaving and fire fighting equipment.

3.1.5 All the ship's personnel concerned should muster at a drill wearing lifejackets properly secured. The lifejackets should continue to be worn during lifeboat drills and launchings but in other cases they may subsequently be removed at the Master's discretion if they would impede or make unduly onerous the ensuing practice, provided they are kept ready to hand.

3.1.6 The timing of emergency drills should vary so that personnel who have not participated in a particular drill may take part in the next.

3.1.7 Any defects or deficiencies revealed during drills and the inspections which accompany them should be made good without delay.

3.2 Fire drills

3.2.1   Efficient fire-fighting demands the full co-operation of personnel in all departments of the ship. A fire drill should be held simultaneously with the first stage of the abandon ship drill. Fire-fighting parties should assemble at their designated stations. Engine room personnel should start the fire pumps in machinery spaces and see that full pressure is put on fire mains. Any emergency pump situated outside machinery spaces should also be started; all members of the crew should know how to start and operate the pump.

3.2.2   The fire parties should be sent from their designated stations to the selected site of the supposed fire, taking with them emergency equipment such as axes and lamps and breathing apparatus. The locations should be changed in successive drills to give practice in differing conditions and in dealing with different types of fire so that accommodation, machinery spaces, store rooms, galleys and cargo holds or areas of high fire hazard are all covered from time to time.

3.2.3   An adequate number of hoses to deal with the asumed fire should be realistically deployed. At some stage in the drill, they should be tested by bringing them into use, firstly with water provided by the machinery space pump and secondly with water from the emergency pump alone.

3.2.4   The drill should extend, where practicable, to the testing and demonstration of the remote controls for ventilating fans, fuel pumps and fuel tank valves and the closing of openings.

3.2.5   Fixed fire extinguishing installations should be tested to the extent practicable.

3.2.6   Portable fire extinguishers should be available for demonstration of the manner of their use. They should include the different types applicable to different kinds of fire. At each drill, one extinguisher or more should be operated by a member of the fire party, a different member on each occasion. Extinguishers so used should be recharged before being returned to their normal location or sufficient spares should otherwise be carried for demonstration purposes.

3.2.7   Breathing apparatus should be worn by members of the fire-fighting parties so each member in turn has experience of its use. Search and rescue exercises should be undertaken in various parts of the ship. The apparatus should be cleaned and verified to be in good order before it is stowed; cylinders of self-contained breathing apparatus should be recharged or sufficient spare cylinders otherwise carried for this purpose.

MS (Closing of Openings in Hulls and Watertight Bulkheads) Regs SI 1987 No. 1298.

3.2.8   Fire appliances, fire and watertight doors and other closing appliances and also fire detection and alarm systems which have not been used in the drill should be inspected to ensure that they are in good order, either at the time of the drill or immediately afterwards. Additionally the relevant statutory requirements should be complied with.

3.3 Survival craft drills

3.3.1   Arrangements for drills should take account of prevailing weather conditions.

3.3.2 Crew members taking part in lifeboat or liferaft drills should muster wearing warm outer clothing and lifejackets properly secured.

3.3.3 Where appropriate, the lowering gear and chocks should be inspected and a check made to ensure that all working parts are well lubricated.

3.3.4 When turning out davits or when bringing boats or rafts inboard under power, seamen should always keep clear of any moving parts.

3.3.5 The engines on motor lifeboats should be started and run ahead and astern. Care should be taken to avoid overheating the engine and the propeller shaft stern gland. All personnel should be familiar with the engine starting procedure.

3.3.6 Hand-operated mechanical propelling gear, if any, should be examined and similarly tested.

3.3.7 Radio equipment should be examined and tested, with the aerial erected, by the Radio Officer or another trained person and the crew instructed in its use.

3.3.8 Water spray systems, where fitted, should be tested in accordance with the lifeboat manufacturer's instructions.

3.3.9 When a drill is held in port, as many as possible of the lifeboats should be cleared and swung out. Each lifeboat should be launched and manoeuvred in the water at least once every three months. Where launching of free-fall lifeboats is impracticable, they may be lowered into the water provided that they are free-fall launched at least once every six months.

3.3.10 When rescue boats are carried which are not also lifeboats they should be launched and manoeuvred in the water every month so far as that is reasonable and practicable. The interval between such drills should not exceed three months.

3.3.11 Where simultaneous off-load/on-load release arrangements are provided great care should be exercised to ensure that the hooks are fully engaged before a boat is recovered, after it has been stowed and prior to launching.

3.3.12 Where davit-launched liferafts are carried then on-board training, including an inflation, must be carried out at intervals not exceeding four months. Great care should be taken to ensure that the hook is properly engaged before taking the weight of the raft. The release mechanism should not be cocked until just prior to the raft landing in the water. If the raft used for the inflation is part of the ship's statutory equipment and not a special training raft, then it MUST be repacked at an approved service station.

3.3.13 Where the handle of the lifeboat winch would rotate during the operation of the winch, it should be removed before the boat is lowered on the brake or raised with an electric motor. If a handle cannot be removed, personnel should keep well clear of it.

3.3.14 Personnel in a rescue boat or survival craft being lowered should remain seated, keeping their hands inside the gunwale to avoid them being crushed against the ship's side. Lifejackets should be worn. In totally enclosed lifeboats seat belts should be secured. Only the launching crew should remain in a lifeboat being raised.

3.3.15   During drills, lifebuoys and lines should be readily available at the point of embarkation.

3.3.16   While craft are in the water, crews should practice manoeuvring the vessel by oar, sail or power as appropriate and should operate the water spray system where fitted on enclosed lifeboats.

3.3.17   Seamen should keep their fingers clear of the long-link when unhooking or securing blocks on to lifting hooks while the boat is in the water, and particularly if there is a swell.

3.3.18   Before craft in gravity davits are recovered by power, the operation of the limit switches or similar devices should be checked.

3.3.19   A portable hoist unit used to recover a craft should be provided with a crutch or have an attachment to resist the torque. These should be checked. If neither device is available, the craft should be raised by hand.

3.3.20   Where liferafts are carried, instruction should be given to the ship's personnel in their launching, handling and operation. Methods of boarding them and the disposition of equipment and stores on them should be explained.

3.3.21   The statutory scale of lifesaving appliances must be maintained at all times. If the use of a liferaft for practice would bring equipment below the specified scale, a replacement must first be made available.

3.4   Action in the event of fire

3.4.1   The risk of fire breaking out on board a ship cannot be eliminated but will be much reduced if the advice given elsewhere in the Code is conscientiously followed at all times (see Chapter 2).

3.4.2   Training in fire-fighting procedures and maintenance of equipment should be assured by regular drills in accordance with section 3.2, but it is important also that access to fire-fighting equipment should be kept unimpeded at all times and that emergency escapes and passage ways are never obstructed.

3.4.3   A fire in its first few minutes can usually be readily extinguished; prompt and correct action is essential.

3.4.4   If fire breaks out, the alarm should be raised and the bridge informed immediately. If the ship is in port, the local fire authority should be called. If possible, an attempt should be made to extinguish or limit the fire, by any appropriate means readily available, either using suitable portable extinguishers or by smothering the fire as in the instance of a fat or oil fire in a galley.

3.4.5   The different types of portable fire extinguishers on board are appropriate to different kinds of fire. Water extinguishers should not be used on oil or electrical fires.

3.4.6   Openings to the space should be shut to reduce the supply of air to the fire and to prevent it spreading. Any fuel lines feeding the fire or threatened by it should be isolated. If practicable combustible materials adjacent to the fire should be removed.

3.4.7 If a space is filling with smoke and fumes, any personnel not properly equipped with breathing apparatus should get out of the space without delay; if necessary, escape should be effected by crawling on hands and knees because air close to deck level is likely to be relatively clear.

3.4.8 After a fire has been extinguished, precautions should be taken against its spontaneous re-ignition.

3.4.9 Personnel, unless wearing breathing apparatus, should not re-enter a space in which a fire has occurred before it has been fully ventilated.

Chapter 4
Safety Officials

4.1 Introduction and Application

4.1.1 The employer is ultimately responsible for the safety of all persons on board ship. However immediate responsibility for the overall safety of the ship and of those on board rests with the Master. Under him each individual member of the ship's crew has a duty to ensure safety in those matters within his own control, whether supervising or carrying out a task, or in reporting or remedying defects which might impair safety. All the safeguards and other facilities provided for the safety of the seafarer should be used.

4.1.2 The development of the necessary degree of safety consciousness and the achievement of high standards of safety depend on foresight, good organisation and the whole-hearted support of management and of all members of the crew. It is therefore important that arrangements should exist on every ship whereby the ship's complement can co-operate and participate in establishing and maintaining safe working conditions.

4.1.3 There is considerable scope in the shipping industry for reducing the number of deaths and injuries resulting from accidents by improving safety in the everyday working and leisure environment. That should be the prime concern of the safety officials on board ship and it is mainly to that end that the information and guidance in this Chapter has been produced.

MS(Safety Officials and Reporting of Accidents and Dangerous Occurences) Regs SI 1982 No 876 ('SORADO' Regs) Part I.

4.1.4 The term 'safety official' includes Safety Officers, safety representatives and other members of safety committees. Merchant Shipping Regulations lay down requirements for the appointment and duties of ships' Safety Officers (see sections 4.2 and 4.6 of this Code) and safety committee (section 4.4) and for the election of safety representatives with specified powers (section 4.3). These requirements should help to ensure that all company policies reflect a commitment by top management to give seafarers, as far as possible, protection at least equivalent to that given to industry ashore.

'SORADO' Regs Part I

4.1.5 The Regulations apply to a UK registered ship on which a crew of more than 5 is employed unless it is (a) non-sea-going, or (b) a fishing vessel, or (c) an offshore installation on its working station or (d) a pleasure craft (see 4.1.6–4.1.8 below).

4.1.6 A non-sea-going ship (excluding a passenger ship) is one which normally remains within the seaward limit of a harbour or, if it goes outside that limit, returns to its port of departure within 24 hours without having called at any other port. The operative word here is 'normally'. A ship which occasionally visits another port would not be subject to the Regulations; a ship spending a significant proportion of its working time making such voyages would fall within the Regulations.

4.1.7 Pleasure craft means a vessel primarily used for sport or recreation. Any vessel which provides sport or recreation to a fee-paying passenger is not a pleasure craft. This is because the sport and recreation are incidental to the main function of the vessel which is earning an income for its owner. Such a ship is primarily used for business.

'SORADO' Regs, Reg 3(7)

4.1.8 In addition to the general exclusions, Regulation 3(7) permits the Secretary of State to grant ad hoc exemptions to specific ships or classes of ships subject to any relevant special conditions. This is to allow different arrangements to be made in cases where the requirements of the Regulations would be difficult to apply. An example might be a multi-crew ship with alternate crews working on a regular shift basis. In considering any request for exemption, the Department of Transport would require to be satisfied that alternative arrangements existed, and would make it a condition of the exemption that these were continued.

4.1.9 Parts of this Chapter will be found helpful to persons who are not safety officials. For example the advice on investigating accidents (section 4.10) may also be useful on ships which are not required to have Safety Officers under Merchant Shipping legislation (see 4.1.5–6). Likewise officers and ratings in a ship with a Safety Officer can refer (in sections 4.3 and 4.7) to the statutory conditions for electing their safety representatives and the powers which the elected representatives hold.

4.2 Appointment of Safety Officers

Reg 3(1)(a)

4.2.1 The employer, who is defined as the person employing the Master, is required to appoint a Safety Officer on board every ship to which the Regulations apply. The employer may or may not be the owner of the ship.

4.2.2 It is important that the employer chooses the right person for Safety Officer. He is the adviser aboard the ship and can be an invaluable assistant to management in meeting the statutory responsibilities for occupational safety. He should be interested in occupational safety, in undertaking the appointment and should have attended a suitable Safety Officer's training course. If it is decided to appoint an officer as the Safety Officer by virtue of his position on board ship, he is more likely to perform well when management is clearly committed to occupational safety and recognises the importance of the Safety Officer's role and that the Safety Officer's duties have to be performed in addition to another job—usually as a watchkeeping officer.

4.2.3 Although not prohibited by the Regulations the appointment of the Master as the Safety Officer is not generally advisable. This is because the Safety Officer is required amongst his other duties to make representations and recommendations on occupational safety to the Master.

4.2.4 If possible the employer should avoid appointing as Safety Officer anyone to whom the Master delegates the task of giving medical treatment. This is because the Safety Officer is statutorily charged with the task of investigating accidents, and will have investigative functions to perform at the scene of an accident to which he may not be able to give proper attention if he is also ministering to the medical needs of the casualties. The Master must record the appointment of a Safety Officer in the official logbook.

4.3 Election of Safety Representatives

Reg 3(4)

4.3.1 These are elected by the crew and, subject to the Regulations anyone is eligible. The Regulations specify that no safety representative may have less than 2 years consecutive sea service since attaining the age of 18, which in the case of a safety representative on board a tanker shall include at least 6 months service on such a ship.

Reg 8(11)

4.3.2 An employer must make rules for the election of safety representatives and cannot disqualify particular persons. It is recommended that the employer should consult with any seafarers' organisation representing his employees when making these rules for elections. Under these Regulations the Master must organise an election of a safety representative within 3 days of being requested to do so by any 2 persons entitled to vote in such an election.

4.3.3 Every safety representative has a statutory right to be a member of the safety committee which must be formed, if not already in existence, as soon as a safety representative is elected.

4.3.4 The number of safety representatives who can be elected varies according to the size of the crew as follows:

6–15 crew	1 elected by officers and ratings together.
16 + crew	1 elected by the officers and 1 elected by the ratings.
Over 30 ratings	1 elected by the officers and 3 by the ratings (ie one each from the deck, engine-room and catering departments, general purpose ratings being included in the deck department).

4.3.5 The Master must record the election of every safety representative in the official logbook.

4.4 Safety Committees

4.4.1 In all ships to which the Regulations apply, it is desirable for the Master to establish a safety committee. However, a statutory requirement for a safety committee only exists on those ships where safety representatives are elected.

Reg 3(5)

4.4.2 The composition of a safety committee is laid down in the Regulations but this does not preclude the appointment of others as committee members nor the appointment of temporary members from time to time. Whenever possible, a company's shore manager with responsibility for safety should attend safety committee meetings on board ship and should in any event require to see the committee's minutes. On short-haul ferries on which different crews work a shift system a scheme of alternate committee members may be adopted to secure proper representation.

4.4.3 Where large numbers of personnel work in separate departments (eg passenger ship galleys and restaurants), departmental sub-committees should be formed on lines similar to those of the main committee and under the chairmanship of a senior member of the department who should serve as a member of the main safety committee in order to report the views of the sub-committee.

Reg 3(5)

4.4.4 The ship's safety committee should include the Master, the Safety Officer and every safety representative elected in accordance with the rules described in section 4.3.

4.4.5 The Master must record the appointment of a safety committee in the official logbook. He should occupy the position of committee chairman since he has overall responsibility on board for safety and has the necessary authority.

4.4.6 It is preferable to appoint as secretary a person who is not a safety official, as officials need to concentrate on the discussion rather than on recording it.

4.4.7 Other committee members should include the Safety Officer, every safety representative and other persons necessary for the proper conduct of the business (eg Chief Officer, Chief Engineer or Catering Officer). Care should be taken to keep the committee sufficiently compact to maintain interest and to enable it to function efficiently.

4.4.8 The frequency of meetings will be determined by circumstances but a frequency of about every 4–6 weeks should suffice. An interval between meetings of much longer than 6 weeks may suggest inertia and an ineffective safety committee.

4.4.9 An agenda (together with any associated documents and papers, and the minutes of the previous meeting) should be circulated to all committee members in sufficient time to enable them to digest the contents and to undertake any necessary preparatory work before the meeting.

4.4.10 If there is a particularly long agenda, consideration should be given to holding two meetings in fairly quick succession rather than one long marathon. If two meetings are held, priority at the first meeting should, of course, be given to the more important or urgent matters.

4.4.11 The first item on the agenda should always be the minutes of the previous meeting. This allows any corrections to the minutes to be recorded and gives the opportunity to report any follow-up action taken.

4.4.12 The last item but one should be any other business. This enables last minute items to be introduced, and prevents the written agenda being a stop on discussion.

4.4.13 The last item on the agenda should be the date, time and place of the next meeting.

4.4.14 Minutes of each meeting should record concisely the business discussed and conclusions reached. A copy should be provided to each committee member. Normally, they should be agreed as a true record at the next meeting, or amended if necessary, under the first item of the agenda (see 4.4.11).

4.4.15 A minutes file or book should be maintained, together with a summary of recommendations recording conclusions reached, in order to provide a permanent source of reference and so ensuring continuity should there be changes in the personnel serving on the committee.

4.4.16 The ship's complement should be kept informed on matters of interest which have been discussed by summaries or extracts from the minutes posted on the ship's notice board. Suggestions may be stimulated by similarly posting the agenda in advance of meetings.

4.4.17 Relevant extracts of agreed minutes should be forwarded through the Master to the Company even though certain matters there-in may have been already taken up with them.

4.5 Termination of Appointments

Reg 4(1) 4.5.1 A Safety Officer's appointment terminates as soon as he ceases to be employed in the particular ship or the employer terminates the particular appointment. The Regulations make no provisions for a Safety Officer to resign his appointment. It is, however, to be hoped that an employer would not continue an appointment just because the person concerned occupied a particular position on the ship, especially if there were a more suitable and perhaps willing person available.

4.5.2 **A safety representative** cannot have his appointment terminated by the employer or Master. He can resign or the crew can elect another in his place. Otherwise he remains a safety representative for as long as he serves on the ship.

4.5.3 **A safety committee** may be disbanded only when there is no longer an elected safety representative on board. A safety committee can, however, operate whether or not there is an elected safety representative.

4.6 Advice to Safety Officer

Reg 5(1) 4.6.1 The Safety Officer is required by the Regulations to try to ensure compliance with the provisions of this Code and of the employer's occupational health and safety policy; and to investigate notifiable accidents to persons on board ship or during access, as well as every dangerous occurance and all potential hazards to occupational health and safety, and to make recommendations to the Master. He also has other specific statutory duties which are listed in paragraphs 4.6.2 to 4.6.9 below.

Reg 5(2) 4.6.2 The Safety Officer is required to investigate all complaints by crew members about occupational health and safety unless he has reason to believe that a complaint is of a frivolous or vexatious nature.

Reg 5(3) 4.6.3 The Regulations require him to carry out occupational health and safety inspections of each accessible part of the ship at least once every three months, or more frequently if there have been substantial changes in the conditions of work. For guidance on this duty see paragraph 4.6.14 et seq.

Reg 5(4) 4.6.4 He also has to make representations and, where appropriate, recommendations to the Master and through him to the employer about any deficiency in the ship in respect of legislative requirements relating to occupational health and safety, relevant Merchant Shipping Notices and the provisions of this Code.

Reg 5(5) 4.6.5 The Safety Officer is required to ensure that safety instructions, rules and guidance are complied with. These include the requirements and guidance referred to in the previous paragraph.

Reg 5(6) 4.6.6 He is required to maintain a record book describing all the circumstances and detail of all accidents and dangerous occurrences (see 4.10.13–15), and details of other action described in paragraph 4.6.2-4 inclusive.

Reg 5(7) 4.6.7 The Safety Officer is required to make the records referred to in paragraph 4.6.6 available on request to any safety representative, to the safety committee, to the Master and to the Department of Transport.

Ref 5(8)

4.6.8 The Safety Officer is also required to stop any work which he reasonably believes may cause a serious accident and immediately to inform the Master (or his deputy) who is responsible for deciding when work can safely be resumed.

Reg 5(9)

4.6.9 Finally, he has a duty to carry out any occupational health or safety investigations or inspections required by the safety committee.

4.6.10 However the Safety Officer is not required by these Regulations to take any of the actions described in paragraphs 4.6.1 to 4.6.9 at a time when emergency action to safeguard life or the ship is being taken.

4.6.11 In carrying out the statutory duties described in paragraphs 4.6.1–9 the Safety Officer should be on the lookout for any potential hazards and the means of preventing accidents. He should try to develop and sustain a high level of safety consciousness among the crew so that individuals work and react instinctively in a safe manner and have full regard to the safety not only of themselves but also of others. He should aim to become the ship's adviser on occupational safety to whom the Master, officers and ratings alike will naturally turn for advice or help on safe working procedures aboard ship.

4.6.12 Example is of prime importance and must be set from the top. If a Safety Officer feels that a ship's officer is not setting a good example a direct approach to that officer suggesting that he mend his ways is often the best course of action. If this fails or is not considered to be appropriate then the Safety Officer might use the safety committee to raise occupational safety as a general topic using examples of dangerous or unsafe practices in the area of the officer concerned. As a last resort the Safety Officer might consider an approach to the Master to use his influence with the officer concerned.

4.6.13 It is essential that everyone joining the ship understands the safety regime from the start. Although the employer should have provided each new employee with a copy of the company's safety policy, the Safety Officer should satisfy himself that each new entrant is informed as soon as possible after he boards, either by his supervisor or by the Safety Officer, of the occupational safety arrangements and the importance attached to them. The induction should include an introduction to the various departments of the ship in company with the Safety Officer or other responsible person, during which particular hazards could be pointed out. Finally, the Safety Officer should ensure that a responsible officer or Petty Officer has, wherever possible, made arrangements for a young new entrant to work with a crew member who is himself thoroughly safety conscious and preferably willing to teach safety at the same time as he displays it. Every effort should be made to see that a new entrant does not work with a person whose attitude to safety is casual or slap-dash. Older hands coming on board for the first time should be reminded of the need to maintain a high level of safety consciousness and of the importance of setting a good example to the less experienced crew member.

4.6.14 Paragraph 4.6.3 explains that the Safety Officer is required to inspect each accessible part of the ship at least every three months, or more frequently if there have been substantial changes in the conditions of work. The Regulations do not define what is meant by 'accessible' or 'substantial changes in the conditions of work'. However, for practical purposes 'accessible' should be taken as meaning all those parts of the ship to which any member of the crew has access without prior authority. Deciding whether 'substantial changes in the conditions of work' have taken place is a matter of judgement. Changes are not limited to physical matters such as new machinery but can also include changes in working practices or the presence of possible new hazards. The Safety Officer is required to keep a record of all inspections as shown in 4.10.13–4.10.15.

4.6.15 The Regulations do not require the Safety Officer to carry out a complete inspection of the ship at one time, only that he inspects each accessible part of the ship every 3 months. Sometimes an inspection of the whole ship at one time would take far too long, and could result in a lack of thoroughness and neglect of his other duties. It is also much easier to get quick and effective action on recommendations arising out of an inspection of one section than of the whole ship. When inspecting a section for which an officer or Petty Officer is responsible the Safety Officer should be accompanied by that Officer.

4.6.16 Before an inspection, the safety officer should read the previous reports of inspections of the particular section, the recommendations made and the subsequent action taken. He should note any recurring problems and, in particular, recommendations for action which have not been put into effect. It is important, however, that the safety officer should not allow previous inspections to prejudice a forthcoming inspection.

4.6.17 It is not possible to give a definite checklist of everything to look for but safe access, the environment and working conditions are major items. The guidance in 4.6.18–4.6.21 offers some further suggestions.

4.6.18 The following are examples of questions the Safety Officer should consider in respect of access and safe movement:
— Are means of access, if any, to the area under inspection (particularly ladders and stairs), in a safe condition, well lit and unobstructed?
— If any means of access is in a dangerous condition, for instance when a ladder has been removed, is the danger suitably blocked off and warning notices posted?
— Is access through the area under inspection both for transit and working purposes clearly marked, well lit, unobstructed and safe?
— Are fixtures and fittings over which seamen might trip or which project, particularly overhead, thereby causing potential hazards, suitably painted or marked?
— Is any gear, which has to be stowed within the area, suitably secured?
— Are all guard-rails in place, secure and in good condition?
— Are all openings through which a person could fall, suitably fenced?
— If portable ladders are in use, are they properly secured and at a safe angle?

4.6.19 The following questions are examples relating to the environment:
— Is the area safe to enter?
— Are lighting levels adequate?
— Is the area clear of rubbish, combustible material, spilled oil etc?
— Is ventilation adequate?
— Are members of the crew adequately protected from exposure to noise when necessary?
— Are dangerous goods or substances left unnecessarily in the area or stored in a dangerous manner?
— Are loose tools, stores and similar items left lying around unnecessarily?

4.6.20 On working conditions the following examples may be appropriate:
— Is machinery adequately guarded where necessary?
— Are any necessary safe operating instructions clearly displayed?
— Are any necessary safety signs clearly displayed?
— Are permits-to-work used when necessary?

— Are crew working in the area wearing any necessary protective clothing and equipment?
— Is that protective clothing and equipment in good condition and being correctly used?
— Is there any evidence of defective plant or equipment and if so what is being done about it?
— Is the level of supervision adequate, particularly for inexperienced crew?
— What practicable occupational safety improvements could be made?

4.6.21 Finally here are some other matters the Officer will find relevant:
— Are all statutory regulations and company safety procedures being complied with?
— Is the safety advice in publications such as this Code, Merchant Shipping Notices etc being followed where possible?
— Have the crew in the area any safety suggestions to make?
— Have any faults identified in previous inspections been rectified?

4.6.22 The Safety Officer's role should be a positive one in that he should seek to initiate or develop safety measures before an accident occurs rather than afterwards. In addition to the specific statutory duties referred to in 4.6.1 he should advise the Master on all matters of safe working practice and assist him in the elimination of accidents and injuries on board ship. For example he should provide a channel by which suggestions for improving safety may be transmitted from seagoing personnel to management.

4.6.23 In carrying out the functions referred to above, the Safety Officer should, with the approval of or at the direction of the Master:
(a) Arrange the distribution of booklets, leaflets and similar advisory and informative material concerning safety matters;
(b) supervise the display of posters and notices and their replacement or renewal in due time;
(c) arrange for the showing of films of safety publicity and, where appropriate, organise subsequent discussions on the subjects depicted;
(d) encourage members of the crew to submit ideas and suggestions for improving safety and enlist their support for any proposed safety measures which may affect them (the person making a suggestion should always be informed of decisions reached and any action taken);
(e) consider any other ways of creating and maintaining interest in improving safety;
(f) receive and draw attention as appropriate to relevent shipping legislation, Department of Transport Merchant Shipping Notices and company and ship's rules and instructions relating to safety of work about the ship. Special regard should be had for persons new on the ship and their attention should always be drawn to any special hazards on the ship.

4.6.24 It is very important that the Safety Officer maintains a good relationship with safety representatives and works in close liaison with them. A good Safety Officer will automatically invite the safety representative to join him in inspecting part of the ship or investigating an accident. He will, whenever possible, consult the safety representative and draw him into discussions about occupational safety matters and arrangements. The safety representative should be able to consult the Safety Officer with the minimum of delay and to expect a considered and reasoned response to any representations. The Safety Officer should associate the safety representative with any follow-up action he may take on the basis of the safety representative's recommendations.

4.6.25 The Safety Officer's relationship with the safety committee is rather different since he is both a member of the committee and also to some extent subject to its direction. A committee has the right to inspect any of the records which a safety officer is required by law to keep, and has the power to require the Safety Officer to carry out any occupational health or safety inspections considered necessary. The Safety Officer should not, generally, be appointed secretary to the safety committee as he needs to concentrate upon the discussion and advise the Master.

4.6.26 The Safety Officer, but not the safety representative, has the power to stop any work, which he believes may cause a serious accident. He must immediately report the stoppage to the Master or his deputy who then has the responsibility for deciding if and when the work should recommence and on what conditions. The Safety Officer may not stop emergency action to safeguard life even though that action may itself involve a risk to life.

4.6.27 One of the specific duties laid upon Safety Officers is to make representations to the Master and through him to the employer about deficiencies on the ship in respect of any legislative requirement relating to occupational safety or health. In order to fulfil this function properly the Safety Officer must be fully conversant with the appropriate regulations. The introduction of new regulations or of amendments to existing regulations will be announced in Merchant Shipping Notices issued by the Department of Transport.

4.6.28 A Safety Officer should not be deterred by the possibility of liability for damages in civil cases arising out of injuries suffered as a result of accidents. The duties placed upon him are limited. If a Safety Officer had carried out the required inspections in a reasonable manner and had reported any revealed breaches of occupational safety provisions to the Master, and through him to the employer, the Safety Officer's legal responsibility in regard to any such breaches would be extinguished. The onus for compliance with the statutory provisions would remain with the person on whom the obligation to comply is placed in the relevant statutory instrument. The Department has been advised that no action for damages would be likely to succeed where a Safety Officer can show that he took all reasonable steps to fulfil his duties.

4.7 Advice to Safety Representatives

Reg 6

4.7.1 Unlike the Safety Officer, the safety representative has powers not duties, although membership of the safety committee imposes certain obligations. In other words, he has the power to do certain things if he so wishes but he need not do them unless he does so wish. He has the right either to participate in normal investigations and inspections made by the safety officer or to undertake similar activities himself. He is also entitled to request, through the safety committee, that the Safety Officer undertakes an investigation and he is entitled to inspect any of the records the Safety Officer is required to keep under the Regulations. He should ensure that he sees the report submitted to the Department of Transport in respect of

'SORADO' Regs Part II
every accident covered by the Regulations.

4.7.2 The safety representative should be fully conversant with all the occupational safety regulations listed in Appendix 3 to this Code and in the

MSN No M1398
relevant Merchant Shipping Notice.

4.7.3 The effectiveness of the safety representative will depend to a large extent upon the relationship which develops between himself and the employer, the Master and Safety Officer. A cooperative spirit is essential, and a good safety representative will try to assist and, where necessary, give encouragement to the Safety Officer in his duties. He will not try to supplant the Safety Officer. He will, however, endeavour to encourage those he represents to become more safety conscious in their various shipboard tasks.

4.7.4 The safety representative should always put forward his views and recommendations in a firm but reasonable and helpful manner. He should be sure of the facts, and be aware of the legal position and of what is practicable. He should request to be kept informed of actions stemming from his recommendations, or the reasons why action was not possible.

4.7.5 If the safety representative should find himself in a situation where his efforts are obstructed, or he is denied facilities, then he should bring the matter to the attention of the Safety Officer or to the Master through the safety committee.

4.7.6 It should be the aim to settle any contentious issue on board ship or thereafter with the employer. If this proves impossible the problem should be referred to a trade union or to the Department of Transport.

4.8 Advice to Safety Committees

'SORADO' Regs, Reg 7

4.8.1 The safety committee is the forum on board ship in which the Master and the appointed and elected safety officials and others meet to discuss matters relating to occupational safety. Its effectiveness will depend upon the willingness of its members, in particular the Master, to give the necessary time and interest to its meetings. In a sense it is the most important element in the occupational safety organisation on board since its membership contains all the safety officials. With the master as its chairman, the committee has the means to take effective action in all matters which it discusses other than those requiring the authorisation of the employer.

4.8.2 The Regulations make no provision about the organisation of the work of the committee but in 4.4.8–4.4.17 some suggestions are offered about frequency of meetings, agenda, etc.

4.9 Advice to the Employer and Master

Reg 8

4.9.1 The effective functioning of the safety organisation on a ship will depend greatly upon the support of the employer and Master. In their own interests the employer and Master are therefore urged to interpret liberally their duties which are set out in the Regulations. These duties are, in general, to facilitate the work of safety officials in carrying out their occupational health and safety functions. For example they are required to provide them access to necessary information, documents etc and inform them of any dangerous cargoes on board and of the dangers which these may cause and of known hazards on board ship. They should also provide the necessary accommodation, permit inspections, allow safety officials necessary training on board, allow them necessary paid absence from ship duties etc. A positive approach to occupational safety is expected from employers and Masters. For example, when any substantial change in personnel occurs the Master should draw the crew's attention to its right to elect safety representatives.

Reg 8(5)

4.9.2 As regards training, the Regulations do not go beyond the requirement to provide Safety Officers and safety representatives with time off from normal duties to undertake any necessary on board training in their respective safety functions. However, as all safety officials have other shipboard duties to perform, any on board training will require very careful planning. Moreover, Safety Officers cannot be expected to fulfil the function of shipboard advisers on occupational safety knowledgeably unless they have previously undertaken a proper training course for the appointment.

Reg 8(8)

4.9.3 An employer's or Master's reaction to representations made to him by the safety officials is crucial, since it will be a very large factor both in determining the level of importance attached to occupational safety and in the relationships between the employer and Master and safety officials. All representations received, from whichever source, should be considered carefully. If there is likely to be a delay in giving an answer, then whoever has made the representations should be informed as soon as possible. Safety suggestions should be implemented whenever it is feasible and reasonable to do so and as quickly as possible. Although employers are required only to specify in writing the reasons for refusing to implement any occupational health or safety recommendations, it is nevertheless a good idea to express thanks for all such suggestions.

4.9.4 It is most important that the Master takes a close interest in the work of the safety officials on board. He needs to encourage the Safety Officer by showing interest and by asking questions; the latter also acts as a check that the Safety Officer is fulfilling his duties effectively. As chairman of the safety committee the Master is in much the best position to ensure that the committee works successfully. Particular attention should be given to the encouragement of any safety representative. Every effort should be made to convince the safety representative that he is a member of a team working for a common end.

4.9.5 Managers should also ensure that all seafarers in the ships under their supervision are provided with information on matters affecting their health and safety at work and, in addition, should make available to appointed Safety Officers and safety committees such further information which they need to enable them to carry out their functions properly. Such information should include that of a technical nature about the hazards and precautions deemed necessary to eliminate or minimise them, in respect of machinery, plant, equipment, processes, systems of work and substances in use at work, or carried as cargo (where these matters are not already covered by official regulations), including relevant information provided by the designer, manufacturer, importer or supplier of any article or substance used by their employees, or carried on their ships.

4.9.6 Where an accident or dangerous occurrence has been reported by a Master, consideration should be given to warning other ships of the occurrence together with appropriate recommendations on action to be taken.

4.9.7 **Security proviso.** This permits the employer to refuse to disclose any information which has been given a national security classification. It does not permit refusal to disclose any other relevant information. Normally the security proviso will apply only on board Royal Fleet Auxiliary vessels but it may apply on occasions on other vessels.

4.10 **Accident Investigation**

4.10.1 The investigation of accidents play a very important part in occupational safety. It is by the identification and study of accidents through the Department of Transport's accident reporting system that similar events may be prevented in future.

'SORADO' Regs
Part II
MSN No. M1383

4.10.2 The statutory requirements regarding accident reporting are set out in the Regulations and further information on reporting procedures by Merchant Shipping Notices.

4.10.3 The Master is responsible for the statutory reporting of accidents. Where a Safety Officer is on board, however, it is his statutory duty to investigate every accident and it is expected that the Master will rely extensively on the results and record of the Safety Officer's investigation when completing his report. The various stages of the typical investigation might proceed as follows.

4.10.4 When an accident occurs priority must be given to the safety of the injured and of those assisting them, and to the immediate safety of the area. When sufficient help is available, however, the Safety Officer should, if possible, avoid involvement with the rescue operation and concentrate on establishing the immediate facts concerning the incident.

4.10.5 First he should record the names—and addresses in case of non-crew personnel—of all those present in the vicinity of the accident. All are unlikely to be witnesses to the actual accident but this can be ascertained later. He should then note and mark the position of the injured, and the use and condition of any protective clothing or equipment or of any tools etc likely to have been in use. Possession should be taken of any portable items which might have some relevance to the investigation. Sketches and photographs can be useful.

4.10.6 When the injured have been removed, the Safety Officer should carry out a more detailed examination at the scene of the accident, watching out for any changes which might have occurred since the accident and any remaining hazards.

4.10.7 The points to look out for will depend on the circumstances of the accident. For example after an accident during access the following should be noted:
— the type of access equipment in use;
— the origin of the access equipment, eg ship's own, provided from shore etc;
— the condition of the access equipment itself, noting particularly any damage such as broken guard-rail, rung etc, the position and extent of any damage (so that it may be compared with witnesses' statements), whether the damage was present before the accident, occurred during it or as a result of it (if the damage was present before the accident it might have been potentially dangerous but it may not necessarily have been a factor in the particular accident);
— any effect of external factors on the condition of the equipment, eg ice, water or oil on the surface;
— the deployment of the equipment, ie the location of the quayside and shipboard ends of the equipment;
— the rigging of the equipment, the method of securing, the approximate angle of inclination;
— the use of ancillary equipment (safety net, lifebuoy and lifeline, lighting);
— the safety of shipboard and quayside approaches to the equipment, eg adequate guard-rails, obstructions and obstacles etc;
— any indication of how the accident might have happened, but remember that subsequent interviews with witnesses must be approached with an open mind;

— weather conditions;
— distances where these are likely to be helpful or relevant.

4.10.8 Interviews of witnesses should take place as soon as possible after the accident when memories are still fresh. Do not overlook those who were not actually witnesses but who may nevertheless have valuable contributions to make, for example a crewman who was present when an order was given. If it is not possible for some reason to interview a particular person, he should be asked to send to the safety officer his own account of the incident.

4.10.9 The actual interview should be carried out in an informal atmosphere designed to put the witness at his ease. Start by explaining the purpose of the interview and obtain some details of the witness's background. Keep any personal bias out of the interview. Ask the witness to relate the event in his own way with as few interruptions as possible; and then test the accuracy of what you have been told. There may, for example, be discrepancies between the account of one witness and those of other witnesses, between different parts of a statement, or with your own observations, which you may want to query. Avoid leading questions implying an answer and simple questions requiring only a yes/no answer which save the witness from thinking about what he is saying. Finally, go over the statement with the witness to ensure that it has been accurately recorded.

4.10.10 Statements for signature by the witness should be prepared as quickly as possible but, if the witness changes his mind about signing a statement, it should be annotated by the Safety Officer that it has been prepared on the basis of an interview with the witness who had subsequently refused to sign it or comment further. Where the witness asks for extensive alterations to the original statement a fresh statement may have to be prepared, but the original one should be annotated by the Safety Officer and retained.

4.10.11 It is helpful to adopt a standard format for statements by accident witnesses. The following is suggested:

VOLUNTARY STATEMENT

Relating to an accident on board/Name of ship/official
number . . . . on/date of accident/at/time of accident.

Particulars of witness:

Name:

Rank and occupation:

Home address of crew members:

Address of employment of others:

STATEMENT OF WITNESS

I make this statement voluntarily having read it before signing it and believing the same to be true.

Signature of Witness

Date   Time

Particulars of interviewer

Name:

Rank:

4.10.12   It is worth emphasising the importance of distinguishing between facts and opinions. Facts can normally be supported by evidence whereas opinions are personal beliefs. An investigation must depend on the facts gathered but opinions can be helpful in pursuing a particular line of enquiry and should not be disregarded.

'SORADO' Regs
Reg 5(6)

4.10.13.   The Safety Officer is required to maintain a record book containing among other things the details of all accidents and the results of their investigation. The Regulations require the Record Book to contain at least the following information:
— details of accidents/dangerous occurrences/investigations/complaints/inspections
— date
— persons involved
— nature of injuries suffered
— all statements made by witness
— any recommendations/representations
— any action taken

4.10.14   Additionally it is suggested that the book should contain the following:
— list of witnesses, addresses, positions and occupations
— whereabouts of original signed statement made by witnesses
— date accident/dangerous occurrence reports sent to Department of Transport if applicable
— list of items collected, why and where stored
— index

4.10.15   The record book should be kept with the ship since it must be made available on request to the safety representative and safety committee, if any. It is also a necessary item of reference for Safety Officers on board the ship. If the ship is sold, and it remains on the UK register, the record book should be transferred with the ship. Where the ship becomes a foreign ship the record book should be retained by the original owners.

CHAPTER 5

Protective Clothing and Equipment

5.1 General

MS (Protective Clothing and Equipment) Regs SI 1985 No 1664

5.1.1 Merchant Shipping Regulations require employers to ensure that every employee engaged in a specified work process, or who may be at risk from such a process, is supplied with suitable protective clothing and equipment. Overalls, gloves and suitable footwear are the proper working dress for most work about the ship but these may not give adequate protection against particular hazards in particular jobs. Details of the protective clothing and equipment required for certain specific work processes are listed in a Merchant Shipping Notice, together with the full title of each relevant standard.

MSN No M1195 (amended by M1358)

5.1.2 Specific recommendations for the use of special protective clothing and equipment will also be found in certain sections of the Code but there will be other occasions when the need for such special protection can only be determined at the time by the officer in charge of the particular operation.

5.1.3 Protective clothing or equipment does nothing to reduce the hazard, it merely sets up a frail barrier against it. The first step in injury prevention should be the elimination of the hazard to the extent that is reasonable and practicable. Personal protective clothing and equipment should be relied upon to afford protection against the hazards that remain.

5.1.4 Defective or ineffective protective equipment provide no defence. It is therefore essential that the correct items of equipment are selected and that they are properly maintained at all times. The manufacturer's instructions should be kept safe with the relevant apparatus and when necessary referred to before use and when maintenance is carried out. The equipment should be kept clean and should be disinfected as and when necessary for health reasons.

5.1.5 A responsible officer should inspect each item of protective equipment at regular intervals and in all cases before and after use. He should ensure that it is returned and properly stowed in a safe place. Personal protective clothing and equipment should always be checked by the wearer each time before use.

5.1.6 All personnel who may be required to use protective equipment should be properly trained in its use and advised of its limitations.

5.1.7 Personal protective clothing and equipment can be classified as follows: Head protection (safety helmets, hair protection); Hearing protection; Face and eye protection (goggles and spectacles, facial shields); Respiratory protective equipment (dust masks, respirators, breathing apparatus); Hand and foot protection (gloves, safety boots and shoes); Body protection (safety suits, safety belts, harnesses, aprons); Protection against drowning (lifejackets, buoyancy aids and lifebuoys).

5.2 Head protection

Safety helmets

5.2.1 Objects falling from a height present a hazard against which safety helmets are most commonly provided. Other hazards include abnormal heat, risk of a sideways blow or crushing, or chemical splashes. These four different types of common risk are given as a guide only and are not intended to be comprehensive.

5.2.2 Since the hazards are so varied in type it will be appreciated that no one type of helmet would be ideal as protection in every case. Design details are normally decided by the manufacturer whose primary consideration will be compliance with an appropriate standard (see paragraph 5.1.1).

5.2.3 The shell of a helmet should be of one piece seamless construction designed to resist impact. The harness or suspension when properly adjusted forms a cradle for supporting the protector on the wearer's head. The crown straps help absorb the force of impact. They are designed to permit a clearance of approximately 25 mm between the shell and the skull of the wearer. The harness or suspension should be properly adjusted before a helmet is worn.

Bump caps

5.2.4 A bump cap is simply an ordinary cap with a hard penetration-resistant shell. They are useful as a protection against bruising and abrasion when working in confined spaces such as a main engine crankcase or a double bottom tank. They do not, however, afford the same protection as safety helmets and are intended only to protect against minor knocks.

Hair nets and safety caps

5.2.5 Personnel working on or near to moving machinery have always to be on their guard against the possibility of loose clothing, jewellery, or their hair becoming entangled in the machinery. In the case of long hair, hair nets or safety caps should be worn where any risk of entanglement exists.

5.3 Hearing protection

5.3.1 All persons exposed to high levels of noise, eg in machinery spaces, should wear ear protectors of a type recommended as suitable for the particular circumstances. Protectors are of three types—ear plugs, disposable or permanent, and ear muffs. For further information see the Code of Practice *Noise Levels in Ships*, published by the Department of Transport.

5.3.2 The simplest form of ear protection is the glass-down ear plug. This type however has the disadvantage of limited capability of noise level reduction. Ear plugs of rubber or plastic also have only limited effect, in that extremes of high or low frequency cause the plug to vibrate in the ear canal causing a consequential loss in protection.

5.3.3 In general, ear muffs provide a more effective form of hearing protection. They consist of a pair of rigid cups designed to completely envelope the ears, fitted with soft sealing rings to fit closely against the head around the ears. The ear cups are connected by a spring loaded headband (or neck band) which ensures that the sound seals around the ears are maintained. Different types are available and provision should be made according to the circumstances of use and expert advice.

5.4 Face and eye protection

5.4.1 In selecting eye and combined eye and face protectors, careful consideration should be given to the kind and degree of the hazard, and the degree of protection and comfort afforded.

5.4.2 The main causes of eye injury are:
(a) infra-red rays—gas welding;
(b) ultra-violet rays—electric welding;
(c) exposure to chemicals;
(d) exposure to particles and foreign bodies.
Protectors are available in a wide variety, designed to British Standard specifications, to protect against these different types of hazard (see 5.1.1).

5.4.3 Ordinary prescription (corrective) spectacles, unless manufactured to a safety standard, do not afford protection. Certain box-type goggles are designed so that they can be worn over ordinary spectacles.

5.5 Respiratory protective equipment

5.5.1 Respiratory protective equipment of the appropriate type is essential for protection when work has to be done in conditions of irritating, dangerous or poisonous dust, fumes or gases. The equipment may be either a respirator, which filters the air before it is breathed, or breathing apparatus which supplies air or oxygen from an uncontaminated source. The selection of the correct respiratory protective equipment for any given situation requires consideration of the nature of the hazard, the severity of the hazard, work requirements and conditions, and the characteristics and limitations of available equipment. Advice on selection and the use and maintenance of the equipment is contained in the relevant British Standard, which should be available to all those concerned with the use of respiratory protective equipment on board ship (see 5.1.1).

BS4275: 1974

5.5.2 It is most important that the face-piece incorporated in respirators and breathing apparatus is fitted correctly to prevent leakage. The wearing of spectacles, unless adequately designed for the purpose, or of beards and whiskers is likely to adversely affect the face seal.

Respirators
5.5.3 The respirator selected must be of a type designed to protect against the hazards being met.

5.5.4 The most common type is the dust respirator, affording protection against dusts and aerosol sprays but not against gases. There are many types of dust respirator available but they are generally of the ori-nasal type, ie half-masks covering the nose and mouth. Many types of light, simple face masks are also available and are extremely useful for protecting against dust nuisance and non-toxic sprays but should never be used in place of proper protection against harmful dusts or sprays.

5.5.5 The positive pressure powered dust respirator incorporates a face-piece connected by a tube to a battery-powered blower unit carried by the wearer to create a positive pressure in the face-piece and thus make breathing easier and reduce face-seal leakage.

5.5.6 The cartridge-type of respirator consists of a full face-piece or half mask connected to a replaceable cartridge containing absorbent or adsorbent material and a particulate filter. It is designed to provide protection against low concentrations of certain relatively non-toxic gases and vapours.

5.5.7   The canister-type of respirator incorporates a full face-piece connected to an absorbent or adsorbent material contained in a replaceable canister carried in a sling on the back or side of the wearer. This type gives considerably more protection than the cartridge type.

5.5.8   The filters, canisters and cartridges incorporated in respirators are designed to provide protection against certain specified dusts or gases. Different types are available to provide protection against different hazards and it is therefore important that the appropriate type is selected for the particular circumstances or conditions being encountered. It must be remembered, however, that they have a limited effective life and must be replaced or renewed at intervals in accordance with manufacturers' instructions.

5.5.9   Respirators provide NO protection against oxygen deficient atmosphere. They should never be used to provide protection in confined spaces such as tanks, cofferdams, double bottoms or other similar spaces against dangerous fumes, gases or vapours. Only breathing apparatus (self-contained or airline) is capable of giving protection in such circumstances (see Chapter 10).

Breathing apparatus
5.5.10   The type of breathing apparatus to be used when entering a space that is known to be, or suspected of being deficient in oxygen or containing toxic gases or vapours is given in section 10.12.

5.5.11   Breathing apparatus should not be used under water unless the equipment is suitable for the purpose, and then only in an emergency.

Resuscitators
5.5.12   It is recommended that resuscitators of an appropriate kind should be provided when any person may be required to enter a dangerous space; see Chapter 10.

5.6   Hand and foot protection

Gloves
5.6.1   The correct type of gloves should be chosen according to the hazard being faced and the kind of work being undertaken. For example, leather gloves are generally best when handling rough or sharp objects, heat-resistant gloves when handling hot objects, and rubber, synthetic or PVC gloves when handling acids, alkalis, various types of oils, solvents and chemicals in general. The exact type selected will depend upon the particular substance being handled, and in these cases expert advice should be followed (see also 1.3.6).

Footwear
5.6.2   Foot injuries most often result from the wearing of unsuitable footwear rather than from failure to wear safety shoes or boots. It is nevertheless strongly advisable that all personnel whilst at work on board ship, wear appropriate safety footwear.

5.6.3   The hazards commonly encountered cause injury as a result of impact, penetration through the sole, slipping, heat and crushing. Safety footwear is available which is designed to protect against these or other specific hazards, manufactured to various British Standards appropriate to the particular danger involved (see 5.1.1).

5.7 Protection from falls

5.7.1 All seamen who are working aloft, outboard or below decks or in any other area where there is a risk of falling more than 2 metres, should wear a safety harness (or belt with shock absorber) attached to a lifeline. Likewise if a vessel is shipping frequent seas, persons on deck should wear a harness and, where practicable, should be secured by lifeline as protection from falls and from being washed overboard or against the ship's structure.

5.7.2 Inertial clamp devices allow more freedom in movement.

5.8 Body protection

5.8.1 Special outerwear may be needed for protection when the seaman is exposed to contact with particular contaminating or corrosive substances. This apparel should be kept for the particular purpose and dealt with as directed in the relevant sections of this Code.

5.9 Protection against drowning

5.9.1 Where work is being carried out overside or in an exposed position where there is a reasonably foreseeable risk of falling or being washed overboard or where work is being carried out in or from a ship's boat a lifebuoy with sufficient line should be provided. In addition and as appropriate a lifejacket or buoyancy aid should be provided.

CHAPTER 6
Signs, Notices and Colour Codes

6.1 General

6.1.1 Colours and symbols appropriately used can provide ever-present information and warnings of hazards which are essential to safety at work, and in some instances may be independent of language. The following provisions are intended to institute a uniform system on ships of United Kingdom registry to the extent that it is practicable to do so, bearing in mind that work is still being carried through in harmonising systems internationally. Those having vision in any way deficient in colour perception should take appropriate care where colour is used as a sole means of identification.

6.2 Signs and Notices

BS 5378: 1980

6.2.1 Safety signs and notices should conform in shape and colour with the relevant British Standard Specification.

6.2.2 Signs of prohibition should be based on a red circular band and red diagonal bar running through the left upper quadrant to the lower right quadrant, with white backing. The symbol for the prohibited action should be shown in black behind the red diagonal bar: for example, 'No Smoking' with a cigarette depicted.

6.2.3 Signs reminding of an essential precaution should comprise a blue disc upon which is superimposed in white a symbol of the precaution to be taken: for example, 'Goggles to be Worn' with a man's head with goggles depicted. If, exceptionally, no suitable symbol is available, appropriate wording may be used instead: for example, 'Keep Clear'.

6.2.4 Warning signs should be based on a yellow triangle bordered by a black band. The symbol for the hazard is depicted in black: for example, poisoning risk with black skull and crossed bones on the yellow background.

6.2.5 Information of a safety nature should be shown by words or a symbol in white upon a green square or rectangle: for example, a white arrow on a green background points to an emergency exit. The same principle applies to fire-fighting equipment and its location except that the background colour should be red.

6.2.6 If there is need to amplify or clarify the meaning of any symbols used in a safety sign or notice, then a supplementary sign with text only (for example, 'Not Drinking Water') should be given below the sign. The supplementary sign should be oblong or square and should either (a) have text in black on a white background or (b) have a background colour which is the same as the safety colour used on the sign it is supplementing, with the text in the relevant contrasting colour.

6.3 Portable fire extinguishers

BS 5423: 1987

6.3.1 Portable fire extinguishers should comply with the relevant British Standard.

6.3.2 The colour of the extinguishers should not conflict with the following recommended systems of colour coding by medium:—

Water	—	Signal Red
Foam	—	Pale Cream
Powder (all types)	—	French Blue
Carbon Dioxide	—	Black
Halon	—	Emerald Green

The area so coded should be large enough to be readily apparent. Where the coding does not cover the whole surface of the extinguisher it is recommended that the remaining area should be either
(a) predominantly signal red, or
(b) of self-coloured (ie natural) metal.

6.4 Electrical wiring

6.4.1 The cores of electrical cables should be identifiable throughout their length by readily identifiable colours or numbers. Although various standards (British, other national and international) exist for colour coding of cores, the colours specified in the standards differ. The colours found on any ship will therefore depend on the country of building or of manufacture of the cables. Care should therefore always be taken to make a positive identification of cable duty, and colours should be used primarily as a means of conductor tracing.

6.4.2 Particular care is required when connecting plugs to domestic equipment which has been brought on to a ship, as a wrong connection may prove fatal. New British equipment should be supplied with cable to the international standard, ie, brown for 'live', blue for 'neutral' and yellow/green for 'earth', but older equipment and that purchased abroad may have other colours.

6.5 Gas cylinders

BS 349: 1973
(See Appendix 1)

6.5.1 Gas cylinders used on United Kingdom ships should be marked and colour coded according to the relevant British Standard Specification.

6.5.2 Each cylinder should be clearly marked with the name of the gas and its chemical formula or symbol. The cylinder body should be coloured according to contents, with, where necessary, a secondary colour band painted around the neck of the cylinder to denote the particular hazards of the gas (flammability, toxicity, etc). Examples of such colour coding on gas cylinders commonly used on board ship are as follows:

Name of gas	Chemical formula or symbol	Ground colour of container	Colour of band
Oxygen	O_2	Black	None
Carbon Dioxide	CO_2	Black	None
Compressed Air	None (mixed gases)	French Grey	None
Nitrogen	N_2	French Grey	Black
Acetylene	C_2H_2	Maroon	None
Propane	None (mixed gases)	Signal Red	None
Butane	None (mixed gases)	None Specified	Signal Red

Note: Cylinders of refrigerant gases are not allocated specified ground or band colours under the British Standard Specification.

BS 1319: 1976
(See Appendix 1)

6.5.3 Medical gas cylinders carried on board should similarly be marked in accordance with the relevant British Standard Specification. The name of the gas or gas mixture contained in the cylinder should be shown on a label affixed to it. The chemical symbol of the gas should be given on the shoulder of the cylinder. The cylinder should also be colour-coded according to contents as shown in the following examples:

Name of gas	Symbol	Colour of body	Colour of valve end
Oxygen	O_2	Black	White
Compressed Air (for breathing app)	AIR	Grey	White and Black

6.6 Pipelines

6.6.1 The following colour coding system is recommended for adoption for the main common pipeline services of United Kingdom registered ships:

Pipe contents	Basic Identification colour	BS Colour reference BS 4800	Colour code band	BS Colour reference BS 4800
Water (Fresh)	Green	12D 45	Blue	18E 53
Water (Salt)	Green	12D 45	None	
Water (Fire Extinguishing)	Green	12D 45	Safety red	04E 53
Compressed Air	Light Blue	20E 51	None	
Steam	Silver Grey	10A 03	None	
Oil (Diesel Fuel)	Brown	06C 39	White	
Oil (Furnace Fuel)	Brown	06C 39	None	
Oil (Lubricating)	Brown	06C 39	Emerald Green	14E 53

BS 1710: 1984

6.6.2 The basic identification colour should be applied on the pipe either over its whole length or as a colour band at regular intervals along the pipe. The colour should similarly be applied at junctions, both sides of valves, service appliances, bulkheads etc, or at any other place where identification might be necessary. Valves on pipelines used for firefighting should be painted red.

6.6.3 Where applicable, the colour code banding should be in approximately 100 mm widths at regular intervals along the length of the pipe on the basic identification colour or painted between two basic identification colour bands each of a width of about 150 mm as shown in the following examples:

Pipe contents	Basic Colour (150 mm approx)	Colour Code (100 mm approx)	Basic Colour (150 mm approx)
Water (Fresh)	Green	Blue	Green
Water (Fire Extinguishing)	Green	Safety Red	Green
Diesel Fuel	Brown	White	Brown

6.6.4 Care should be taken to ensure that when replacing or repainting pipes, valves etc, the correct colour is used.

6.6.5 When it is necessary to know the direction of flow of the fluid, this should be indicated by an arrow situated in the proximity of the basic identification colour and painted white or black in order to contrast clearly with that colour.

6.6.6 Such a system as recommended above would be useful, for instance, in tracing a run of pipes but should not be relied upon as a positive identification of the contents of the pipe; a check should always be made before opening up and precautions taken against the contingency that the content is other than that expected.

6.6.7 Other pipeline systems on ships, such as cargo pipelines, may be colour-coded in a similar fashion but no specific recommendations are made here because a comprehensive system to cover the needs of all types of ship would require so wide a range of colours that contrasts would be small and easily obscured by fading or dirt.

6.6.8 Colour coding of pipelines may vary from ship to ship and seamen moving from one ship to another should ascertain from a competent officer what the colours means on each particular vessel.

6.7 Dangerous goods

MS (Dangerous Goods and Marine Pollutants) Regulations SI 1990 No 2605

6.7.1 All dangerous goods and substances carried as cargo on ships have to be classified, packed and labelled in accordance with Merchant Shipping Regulations.

6.7.2 Examples of the labels to be affixed to packages and containers of dangerous goods, depicting by colour, name and 'hazard diagram' the particular dangers of that substance (flammability, toxicity, corrosiveness etc) are given in the International Maritime Dangerous Goods (IMDG) Code. Labels and other markings are required to be durable enough to remain identifiable on packages surviving at least three months immersion in the sea. Labels manufactured to the relevant British Standard are regarded as meeting the requirement.

BS 5609: 1986

CHAPTER 7

Permit-to-work systems

7.1.1 There are many types of operation on board ship where the routine actions of one man may inadvertently endanger another or when a series of action steps need to be taken to ensure the safety of those engaged in a specific operation. Danger may arise from the activation of a radar installation while men are working in the vicinity of the scanner; unusual risks may arise during the repair and maintenance work when in-built safeguards, effective during normal operation, have to be disturbed; a number of safety measures and precautions need to be taken before work is done in a tank or duct keel.

7.1.2 In all instances it is necessary, before the work is begun, to identify the hazards and then to ensure that they are eliminated or effectively controlled. Ultimate responsibility rests with the employer to see that this is done. Sometimes automatic safeguards on machinery or electrical equipment, for example, may greatly reduce the hazards but normally reliance has to be placed on the people involved following a proper procedure. In those cases verbal instructions, requests and responses which might be misheard, misinterpreted or not fully remembered are not a satisfactory basis for activities in which men's lives may be at risk. A more effective control can be achieved by the use of a written system which requires step by step formal actions by those responsible for the work. Such a system may be instituted by use of a 'permit-to-work'. That essentially is a document which sets out the work to be done, and the precautions to be taken in doing it. It consists basically of an organised and pre-defined safety procedure. It forms a clear record of all the foreseeable hazards which have been considered in advance and the appropriate precautions which have been determined and shows the correct sequence of operations and precautions. A permit-to-work does not in itself make the job safe, but is a guide dependent for its effectiveness upon the conscientious observance of the set procedure by those involved in the job.

7.1.3 The particular circumstances of individual ships will determine the particular areas in which permit-to-work systems can most usefully be adopted but, in general, the following principles should apply:
(a) The first and most important step is the assessment of the situation by a ship's officer who is experienced in the work and is thoroughly familiar with the relevant hazards.
(b) The information given in the permit should be precise, detailed and accurate. It should state exactly the location and details of the work to be done, the nature and results of any preliminary tests undertaken, the measures undertaken to make the job safe and the safeguards that need to be taken during the operation.
(c) The permit should specify the period of its validity (which should not exceed 24 hours) and any time limits applicable to the work which it authorises.
(d) The permit should be recognised as an overriding instruction until it is cancelled.

(e) Only the work specified on the permit should be undertaken.
(f) Before signing the permit, the responsible officer should personally check that all the measures specified as necessary have in fact been taken and that safety arrangements will be maintained until the permit is cancelled.
(g) Anyone who takes over, either as a matter of routine or in an emergency, from the person who originally issued the permit, should assume full responsibility until he has either cancelled the permit or handed it over to another nominated person who should be made fully conversant with the situation.
(h) The person responsible for carrying out the specified work should countersign the permit to indicate his understanding of the safety precautions to be observed. On completion of the work, he should notify the authorising officer.

7.1.4 In many instances a full permit-to-work system would be over-elaborate but there may still be a need to ensure that certain precautions are taken at appropriate stages of the work for the safety of those involved in the work or of those who may be affected by it. A rather simpler check list can be a useful aid in such cases. For example it could cover the posting of warning notices and the isolation of controls where the actuation of machinery or equipment could imperil men working at a place remote from the control position, especially aloft and overside, work on alarm and automatic systems and entry into refrigerated spaces.

7.1.5 The table which follows illustrates a specimen form for a permit-to-work showing the headings that may need to be covered. It can be readily adapted to the exact circumstances of the job to be carried out, by amending wording, by deleting sections not relevant or by other changes which may be suitable.

Specimen of a permit-to-work

Note: The Authorising Officer should indicate the sections applicable by ticks in the left-hand boxes next to headings, deleting any sub-heading not applicable. He should insert the appropriate details when the sections for Other work or Additional precautions are used.

The Authorised Person should tick each applicable right-hand box as he makes his check.

Work to be done (description) **Location** (designation of space,
Authorised person in charge   machinery, etc)
Period of validity of permit   **Crew detailed** (names)
(Should not exceed 24 hours)
Authorising Officer
   (signed)   (time)   (date)

Entry into enclosed or confined spaces

		Checked
1	Space thoroughly ventilated	1
2	Atmosphere tested and found safe	2
3	Rescue and resuscitation equipment available at entrance	3
4	Responsible person in attendance at entrance	4
5	Communication arrangements made between person at entrance and those entering	5

6	Access and illumination adequate	6
7	All equipment to be used is of approved type	7
8	When breathing apparatus is to be used	
	(i) Familiarity of user with apparatus is confirmed	8(i)
	(ii) Apparatus has been tested and found to be satisfactory	8(ii)

Machinery or equipment

1. Removed from service/isolated from sources of power or heat — 1
2. All relevant personnel informed — 2
3. Warning notices displayed — 3

Hot Work

Checked

1. Area clear to dangerous material and gas-free — 1
2. Ventilation adequate — 2
3. Equipment in good order — 3
4. Fire appliances in good order — 4

Other work

Checked

1. — 1
2. — 2
3. — 3
4. — 4

Additional Precautions

1. — 1
2. — 2
3. — 3
4. — 4

Certificate of checks

I am satisfied that all precautions have been taken and that safety arrangements will be maintained for the duration of the work.
Authorised person in charge   (Signed)

Certificate of completion

The work has been completed and all persons under my supervision, materials and equipment have been withdrawn.
Authorised person in charge   (Signed)   (time)   (date)

CHAPTER 8

Means of access

8.1 General

Reg 4 of MS (Means of Access) Regs SI 1988 No 1637

8.1.1 Merchant Shipping Regulations place an obligation on both the Master of a ship and the employer of the master to ensure, so far as is reasonably practicable, that there is a safe means of access between the ship and the shore or another ship. Where the provision of equipment is necessary to achieve safe means of access it must be placed in position promptly, be properly rigged and deployed, safe to use and adjusted as necessary to maintain safety of access. The access equipment and immediate approaches must be adequately illuminated. In carrying out these duties full account must be taken of the principles and the guidance described in this Chapter.

8.1.2 When the access equipment is provided from the shore it is still the responsibility of the Master to ensure as far as is reasonably practicable that the equipment is suitable, properly rigged and deployed, and adequately illuminated.

Reg 10 as above

8.1.3 When suitable access equipment is provided from the ship or from the shore or from another ship, any person boarding or leaving the ship must use that equipment.

8.2 Standards of Construction

Regs 5 and 6 Reg 4 Reg 7

8.2.1 Gangways must be carried on ships of 30 metres or over and accommodation ladders must be carried on ships of 120 metres or over, complying with the specifications in paragraphs 8.2.2, 8.2.3 and 8.2.4 below. Access equipment must be of good construction, sound material and adequate strength, free from patent defect and properly maintained. Rope ladders must comply with the requirements in 8.2.6.

8.2.2 Gangways and accommodation ladders must be clearly marked with the manufacturer's name, the model number the maximum designed angle of use and the maximum safe loading both by number of persons and by total weight.

BSMA 78: 1978

8.2.3 Gangways must comply with the specifications set out in Standard BSMA 78:1978 (excluding the maximum overall widths specified in Table 2), or be of an equivalent standard and must be fitted with suitable fencing along their entire length.

BSMA 89: 1980

8.2.4 Accommodation ladders must comply with the specifications set out in Standard BSMA 89: 1980 or be of an equivalent standard, and must be fitted with suitable fencing along their entire length, except that fencing at the bottom platform may allow access from the outboard side.

8.2.5 When a bulwark ladder is to be used it must comply with the specifications set out in the Shipbuilding Industry Standard No SIS 7, or be of an equivalent standard. Adequate fittings must be provided to enable the bulwark ladder to be properly and safely secured.

8.2.6 A rope ladder must be of adequate width and length and so constructed that it can be efficiently secured to the ship. The steps must provide a slip resistant foothold of not less than 400 mm × 115 mm and must be so secured that they are firmly held against twist, turnover or tilt. The steps must be equally spaced at intervals of 310 mm (± 5 mm). Ladders of more than 1.5 metres in length must be fitted with spreaders not less than 1.8 metres long. The lowest spreader must be on the fifth step from the bottom and the interval between spreaders must not exceed nine steps.

8.3 Maintenance

Reg 4(d)

8.3.1 Any equipment used for the provision of means of access and any safety net must be properly maintained.

8.3.2 All access equipment should be inspected by a competent person at appropriate intervals. Any defects affecting the safety of any access equipment, including access provided by a shore authority, should be reported immediately to a responsible person and should be made good before further use.

8.3.3 No access equipment should be painted or treated in such a way as to conceal any cracks or defects.

8.3.4 Aluminium equipment should be examined for corrosion in accordance with the instructions in Section 8.10.

8.4 Positioning of Access Equipment

8.4.1 The angles of inclination of a gangway or accommodation ladder should be kept within the limits of which it was designed. Gangways should not be used at an angle of inclination greater than 30° from the horizontal and accommodation ladders should not be used at an angle greater than 55° from below the horizontal, unless specifically designed for greater angles.

8.4.2 When the inboard end of the gangway or accommodation ladder rests on or is flush with the top of the bulwark, a bulwark ladder should be provided. Any gap between the bulwark ladder and the gangway or accommodation ladder should be adequately fenced to a height of at least 1 metre.

8.4.3 Gangways and other access equipment should not be rigged on ships' rails unless the rail has been reinforced for the purpose.

8.4.4 The means of access should be checked that it is safe to use after rigging. There should be further checks to ensure that adjustments are made when necessary due to tidal movements or change of trim and freeboard. Guard ropes, chains etc should be kept taut at all times and stanchions should be rigidly secured.

8.4.5 Each end of a gangway or accommodation or other ladder should provide safe access to a safe place or to an auxiliary safe access.

8.4.6 The means of access should be sited clear of the cargo working area and so placed that no suspended load passes over it. Where this is not practicable, access should be supervised at all times.

8.5 Lighting and Safety of Movement

8.5.1 In normal circumstances, the whole means of access and the immediate approaches to it should be effectively illuminated from the ship or the shore to at least a level of 20 lux, as measured at a height of 1 metre above the surface of the means of access or its immediate approaches. Where the dangers of tripping or falling are greater than usual because of bad weather conditions or where the means of access is obscured, eg by the presence of coal dust, consideration should be given to a higher minimum level of say 30 lux.

8.5.2 The means of access and its immediate approaches should be kept free from obstruction and, as far as reasonably practicable, kept clear of any substance likely to cause a person to slip or fall. Where this is not possible, appropriate warning notices should be posted and if necessary the surfaces suitably treated.

8.6 Portable and Rope Ladders

Reg 7

8.6.1 A portable ladder must be used for the purpose of access to the ship only where no safer means of access is reasonably practicable and a rope ladder must be used only for the purpose of access between a ship with high freeboard and a ship with low freeboard or a ship and a boat and where no safer means of access is reasonably practicable.

8.6.2 When it is necessary to use a portable ladder for access it should be used at an angle of between 60° and 75° from the horizontal. The ladder should extend at least 1 metre above the upper landing place unless there are other suitable handholds. It should be properly secured against slipping or shifting sideways or falling and be so placed as to afford a clearance of at least 150 mm behind the rungs.

8.6.3 When a portable ladder is resting against a bulwark or rails, suitable safe access to the deck as recommended in paragraph 8.4.2 should be provided.

8.6.4 A rope ladder should never be secured to rails or to any other means of support unless the rails or support are so constructed and fixed as to take the weight of a man and a ladder with an ample margin of safety.

8.6.5 A rope ladder should be left in such a way that it either hangs fully extended from a securing point or is pulled up completely. It should not be left so that any slack will suddenly pay out when the ladder is used.

8.7 Safety Nets

Reg 9

8.7.1 An adequate number of safety nets of a suitable size and strength are to be carried on the ship or otherwise be readily available. Where there is a risk of a person falling from the access equipment or from the quayside or ship's deck adjacent to the access equipment, a safety net shall be mounted where reasonably practicable. The aim of safety nets is to minimuse the risk of injury arising from falling between the ship and quay or falling onto the quay or deck and as far as reasonably practicable the whole length of the means of access should be covered. Safety nets should be securely rigged, with use being made of attachment points on the quayside where appropriate.

8.8 Life-buoys

Reg 8

8.8.1 A life-buoy with a self-activating light and also a separate buoyant safety line attached to a quoit or some similar device must be provided ready for use at the point of access aboard the ship.

8.9 Special Circumstances and General Guidance

8.9.1 In some circumstances it may not be practicable to mount a proper safe means of access by conventional means (because of, for example, the frequent movement of the ship which is sometimes necessary during loading to facilitate trimming of the cargo, or because of the nature of the loading operation itself). On such occasions access to the vessel should be specially supervised and consideration given to providing alternative means of access, for example, by using shore access arrangements or by using the accommodation ladder on the offshore side of the vessel from which a suitable boat can convey persons safely to and from the shore.

8.9.2 Small boats or tenders used to provide access between the shore and the ship should be safe and stable, be suitably powered, and be properly equipped with the necessary safety equipment and, if not a ship's boat, be approved for that purpose. They should not be used in unsuitable sea conditions.

8.9.3 Where a vessel is moored alongside another vessel, there should be co-operation between the two vessels in order to provide suitable and safe access. Access should generally be provided by the ship lying outboard, except that where there is a great disparity in freeboard access should be provided by the ship with the higher freeboard.

8.9.4 In Ro-Ro and ferry-type ships ramps which are used by vehicles should not be used also for pedestrian access unless there is suitable segregation of vehicles and pedestrians whether by providing a suitably protected walkway or by ensuring that pedestrians and vehicles do not use the ramp at the same time. (See the Department of Transport Code of Practice on the Stowage and Securing of Vehicles on Ro-Ro ships).

8.9.5 Care should be taken at all times when boarding or leaving a ship. Care should be taken also when moving through the dock area, particularly at night. The edges of the docks, quays etc should be avoided and any sign prohibiting entry to an area should be strictly observed. Where there are designated routes they should be followed exactly and this is especially important in the vicinity of container terminals or other areas where rail traffic, straddle carriers or other mechanical handling equipment is operating, since the operators of such equipment have restricted visibility and anyone walking within the working area is at serious risk.

8.10 Corrosion of Accommodation Ladders and Gangways

8.10.1 Aluminium alloys are highly susceptible to galvanic corrosion in a marine atmosphere if they are used in association with dissimilar metals. Great care should be exercised when connecting mild steel fittings, whether or not they are galvanised, to accommodation ladders and gangways constructed of aluminium.

8.10.2 Plugs and joints of neoprene, or other suitable material, should be used between mild steel fittings, washers, etc and aluminium. The plugs or joints should be significantly larger than the fittings or washers.

8.10.3 Repairs using mild steel doublers or bolts made of mild steel or a brass or other unsuitable material should be considered as temporary. Permanent repairs, or the replacement of the means of access, should be undertaken at the earliest opportunity.

8.10.4 The manufacturer's instructions should give guidance on examination and testing of the equipment. However, close examination of certain parts of accommodation ladders and gangways is difficult due to their fittings and attachments. It is essential, therefore, that the fittings are removed periodically for a thorough examination of the parts most likely to be affected by corrosion. Accommodation ladders and gangways should be turned over to allow for a thorough examination of the underside. Particular attention should be paid to the immediate perimeter of the fittings; this area should be tested for corrosion with a wire probe or scribe. Where the corrosion appears to have reduced the thickness of the parent metal to 3 mm, back plates should be fitted inside the stringers of the accommodation ladder or gangways.

CHAPTER 9

Safe movement on board ship

9.1 General

MS (Safe Movement on Board Ship) Regs SI 1988 No 1641

Reg 4

9.1.1 Merchant Shipping Regulations place an obligation on both the Master of a ship and the employer of the Master to ensure that a safe means of access is provided and maintained to any place on the ship at which a person may be expected to be. In carrying out the duties arising from these Regulations full account must be taken of the principles and the guidance contained in this Chapter. Places on the ship at which a person may be expected to be include accommodation areas as well as normal places of work. Persons in this context include passengers, dock-workers, and other visitors to the ship on business but exclude persons who have no right to be on the ship.

9.2 Transit Areas

Reg 5

9.2.1 All deck surfaces used for transit about the ship and all passageways, walkways and stairs shall so far as is reasonably practicable be properly maintained and kept free from materials or substances liable to cause a person to slip or fall. Where necessary for safety, walkways on decks should be delineated by painted lines or otherwise and indicated by pictorial signs. Where normal safe transit across an area is made impracticable, it should be isolated until suitable remedial action can be taken.

9.2.2 Transit areas should where practicable have slip-resistant surfaces. Where an area is made slippery by snow, ice or water, sand or some other suitable substance should be spread over the area. Spillages of oil or grease etc should be cleaned up as soon as practicable.

9.2.3 Gratings in the deck should be properly maintained and kept closed when access to the space below is not required.

9.2.4 Permanent fittings which may cause obstruction and be dangerous to vehicles, lifting appliances or persons should be made conspicuous by means of colouring or marking or lighting. Temporary obstacles can also be dangerous and if they are to be there for some time their presence should also be indicated by appropriate warning signs.

9.2.5 When at sea, any gear or equipment stowed to the side of a passageway or walkway should be securely fixed or lashed against the movement of the ship.

9.2.6 Litter and loose objects, eg tools, should not be left lying around. Wires and ropes should be stowed and coiled so as to cause least obstruction.

9.2.7 When rough weather is expected, life-lines should be rigged securely across open decks.

9.2.8 Particular attention should be given to ensure the safe movement about ship of dock-workers and passengers who will be less familiar with possible hazards, especially on deck.

9.2.9 Shipboard lashing and securing arrangements for deck cargo may call for special measures to ensure safe access to the top of, and across, the cargo.

9.3 Lighting

Reg 6

9.3.1 Those areas of the ship being used for the loading or unloading of cargo, or other work processes or for transit purposes, shall be adequately and appropriately illuminated. Lighting facilities should be properly maintained.

9.3.2 For loading and unloading areas and for other working areas a lighting level of at least 20 lux should be provided and for transit areas a level of at least 8 lux should be provided (both measured at a height of 1 metre above the surface level) unless:—
(a) A higher level is required by other Regulations, eg the Crew Accommodation Regulations, or,
(b) Provision of such levels of lighting would contravene other Regulations, eg the Collision Regulations and the Distress Signals Order.

Where visibility is made worse, eg by fog, clouds of dust, or steam, which could lead to an increase in the risks of accident occurring, the level of lighting should be increased appropriately.

9.3.3 The level of lighting should be such as to enable obvious damage to, or leakage from, packages to be seen. When there is a need to read labels or container plates or to distinguish colours the level of lighting should be adequate to allow this, or other means of illumination should be provided.

9.3.4 Lighting should be reasonably constant and arranged to minimise glare and dazzle, the formation of deep shadows and sharp contrasts in the levels of illumination between one area and another.

9.3.5 Broken or defective lights should be reported to the responsible officer and repaired as soon as practicable.

9.3.6 Before leaving an illuminated area or space a check should be made that there are no other persons remaining within that space before switching off or removing lights.

9.3.7 Unattended openings in the deck should either be kept illuminated or be properly and safely closed before lights are switched off.

9.3.8 When portable or temporary lights are in use, the light supports and leads should be arranged, secured or covered so as to prevent a person tripping, or being hit by moving fittings, or walking into the cables or supports. Any slack in the leads should be coiled. The leads should be kept clear of possible causes of damage eg running gear, moving parts of machinery, equipment and loads. If they pass through doorways the doorways should be secured open. Leads should not pass through doors in watertight bulkheads or fire door openings when the ship is at sea. Portable lights should never be lowered or suspended by their leads.

9.3.9 Where portable or temporary lighting has to be used fittings and leads should be suitable and safe for the intended usage. To avoid risks of electric shock from mains voltage, the portable lamps used in damp or humid conditions should be of low voltage, preferably 12 volts, or other suitable precautions taken.

9.4 Safety Signs

Reg 7 BS 5378 Pt 1 1980

9.4.1 Any safety signs permanently erected on board the ship for the purpose of giving health or safety information or instruction shall comply with the appropriate British Standard or a national or international standard providing for equivalent safety.

9.4.2 Safety signs, which include hazard warnings, should be used whenever a hazard or obstruction exists and such a sign is appropriate. Particular attention should be paid on passenger ships to hazards which may be familiar to seafarers but not to passengers.

9.4.3 Where a language other than English is extensively used on a ship, any text used in conjunction with a sign should usually be displayed also in that language.

9.4.4 Further information on the use of safety signs is contained in Chapter 6 of this Code.

9.5 Guarding of Openings

9.5.1 Any opening, open hatchway or dangerous edge into, through or over which a person may fall shall be fitted with secure guards or fencing of adequate design and construction (see 9.5.3 and 9.5.4 below). These requirements do not apply:—
(a) where any opening affords a permanent means of transit about the ship, to the side of the opening used for access,

Reg 8
(b) where, and to the extent that, the person upon whom a duty is imposed is able to show that the work process being carried out or about to be carried out makes the provision of such guards or fencing not reasonably practicable. This would include short interruptions of work for meals or other purposes.

9.5.2 Any hatchway open for the purposes of handling cargo or stores through which a person may fall should be closed as soon as those operations cease, except during short interruptions of work, including meal breaks, or where closure cannot be effected without prejudice to safety or mechanical efficiency because of the heel or trim of the ship.

9.5.3 The guardrails or fencing should be free from sharp edges and should be properly maintained. Where necessary, locking devices, and suitable stops or toe-boards should be provided. Each course of rails should be kept substantially horizontal and taut throughout their length.

9.5.4 Guardrails or fencing should consist of an upper rail at a height of 1 metre and an intermediate rail at a height of 0.5 metres. The rails may, where necessary, consist of taut wire or taut chain. Where existing fencing to a height of at least 920 mm has been provided this need not be replaced while it remains secure and adequate.

9.6 Ladders

General

Reg 9

9.6.1 All ship's ladders shall be of good construction and sound material, of adequate strength for the purpose for which they are used, free from patent defect and properly maintained.

9.6.2 Where a fixed ladder or stairway is found to have become unsafe or where it has proved necessary to remove such a ladder or stairway, access to that ladder or stairway, or the opening where the ladder or stairway was positioned, should be blocked off and warning notices placed at all approaches.

9.6.3 Suitable hand-holds should be provided at the top and at any intermediate landing place of all fixed ladders.

Hold Access—New Ships

Reg 11

9.6.4 Where the keel of a ship is laid or the ship is at a similar stage of construction, after 31 December 1988 the following standards of hold access shall be provided:—
  (i) The access shall be separate from the hatchway opening, and shall be by a stairway if possible.
  (ii) A fixed ladder, or a line of fixed rungs, shall have no point where they form a reverse slope.
  (iii) The rungs of a fixed ladder shall be at least 300 mm wide, and so shaped or arranged that a person's foot cannot slip off the ends. Rungs shall be evenly spaced at intervals of not more than 300 mm and there shall be at least 150 mm clear space behind each rung.
  (iv) There shall be space outside the stiles of at least 75 mm to allow a person to grip them.
  (v) There shall be a space at least 760 mm wide for the user's body, except that at a hatchway this space may be reduced to a clear space of at least 600 mm by 600 mm.
  (vi) Fixed vertical ladders shall be provided with a safe intermediate landing platform at intervals of not more than 9 metres.
  (vii) Where vertical ladders to lower decks are not in a direct line a safe intermediate landing shall be provided.
  (viii) Intermediate landings shall be of adequate width and afford a secure footing and extend from beneath the foot of the upper ladder to the point of access to the lower ladder. They shall be provided with guard rails.
  (ix) Fixed ladders and stairways giving access to holds shall be so placed as to minimise the risk of damage to them from cargo handling operations.
  (x) Fixed ladders shall, if possible, be so placed or installed as to provide back support for a person using them; but hoops shall be fitted only where they can be protected from damage to them from cargo handling operations.

Hold Access—Existing Ships

9.6.5 Where the keel of a ship was laid or the ship was at a similar stage of construction before 1 January 1989, at least the following standards of hold access should be provided:—
  (i) Access should be provided by steps or ladder, except:
    (a) at coamings; and
    (b) where the provision of a ladder on a bulkhead or in a trunk hatchway is clearly not reasonably practicable.

In such cases ladder cleats or cups may be used.
- (ii) All ladders between lower decks should be in the same line as the ladder from the top deck, unless the position of the lower hatch or hatches prevents this.
- (iii) Cleats or cups should be at least 250 mm wide and so constructed as to prevent a person's foot slipping off the side.
- (iv) Each cleat, cup, step or rung of a ladder should provide a foothold, including any space behind the ladder, at least 115 mm deep. Cargo should not be so stowed as to reduce this foothold.
- (v) Ladders which are reached by cleats or cups on a coaming should not be recessed under the deck more than is reasonably necessary to keep the ladder clear of the hatchway.
- (vi) Shaft tunnels should be equipped with adequate handholds and footholds on each side.
- (vii) All cleats, cups, steps or rungs of ladders should provide adequate handholds.

Portable Ladders

9.6.6 A portable ladder should only be used where no safer means of access is reasonably practicable.

9.6.7 Portable ladders should be pitched between 60° and 75° from the horizontal, properly secured against slipping or shifting sideways and be so placed as to afford a clearance of at least 150 mm behind the rungs. Where practicable the ladder should extend to at least 1 metre above any upper landing place unless there are other suitable handholds.

9.7 Vehicles

9.7.1 Persons selected to drive ships' powered vehicles and powered mobile lifting appliances should be fit to do so, and have been trained for the particular category of vehicle or mobile lifting appliance to be driven, and tested for competence. Persons authorised to operate types of powered vehicles or powered mobile lifting appliances before 1 January 1989 should be considered competent and authorised accordingly.

9.7.2 Authorisations of crew members should either be individually issued in writing or comprise a list of persons authorised to drive. These authorisations may need to be made available for inspection to Dock Authorities.

9.7.3 Maintenance of ships' powered vehicles and powered mobile lifting appliances should be undertaken in accordance with manufacturers' instructions.

9.7.4 Drivers of ships' powered vehicles and powered mobile lifting appliances should exercise extreme care, particularly when reversing.

9.7.5 There should be suitable traffic control arrangements, including speed limits, and where appropriate the use of signallers. Collaboration may be necessary with shore side management where they also control vehicle movements on board ship.

9.7.6 As far as possible routes used by vehicles should be separated from pedestrian passageways.

9.7.7 No ramp used by vehicles should be so steep as to be unsafe.

9.8 Drainage

9.8.1   Decks which need to be washed down frequently or are liable to become wet and slippery, should be provided with effective means of draining water away. Apart from any open deck these places include the galley, the ship's laundry and the washing and toilet accommodation.

9.8.2   Drains and scuppers should be regularly inspected and properly maintained.

9.8.3   Where drainage is by way of channels in the deck, these should be suitably covered.

9.8.4   Duck boards, where used, should be soundly constructed and designed and maintained so as to prevent accidental tripping.

9.9 Watertight Doors

9.9.1   All members of the crew who would have occasion to use any watertight doors should be instructed in their safe operation.

9.9.2   Particular care should be taken when using power operated watertight doors which have been closed from the bridge. If opened locally under these circumstances the door will re-close automatically with a force sufficient to crush anyone in its path as soon as the local control has been released. The local controls are positioned on each side of the door so that a person passing through may open the door and then reach to the other control to keep the door in the open position until transit is complete. As both hands are required to operate the controls, no person should attempt to carry any load through the door unassisted.

9.9.3   Notices clearly stating the method of operation of the local controls should be prominently displayed on both sides of each watertight door.

9.9.4   No-one should attempt to pass through a watertight door when it is closing and/or the warning bell is sounding.

9.10 General Advice to Seafarers

9.10.1   Seafarers are reminded to take care as they move about the ship and to do so in a seamanlike fashion. In particular, the following points, though obvious, are too often overlooked:
(a) the possibility of an unusual lurch or heavy roll of the ship should always be borne in mind;
(b) suitable footwear should be worn which will protect toes against accidental stubbing and falling loads and will afford a good hold on deck and give firm support when using ladders; extra care should be taken when using ladders whilst wearing sea boots or gloves;
(c) it is dangerous to swing on or vault over stair rails, guardrails or pipes;
(d) injuries often happen due to jumping off hatches etc or by stumbling over door sills or other obstacles.

9.10.2   A seafarer who finds any defects in any equipment, or a condition he believes to be a hazard or unsafe, should immediately report it to a responsible person, who should take appropriate action.

CHAPTER 10

Entering Enclosed or Confined Spaces

10.1 General

10.1.1 The atmosphere of any enclosed or confined space may put at risk the health or life of any person entering it. It may be deficient in oxygen and/or contain flammable or toxic fumes, gases or vapours. Such an unsafe atmosphere may be present or arise subsequently in any enclosed or confined space including cargo holds, double bottoms, cargo tanks, pump rooms, compressor rooms, fuel tanks, ballast tanks, cofferdams, void spaces, duct keels, inter-barrier spaces, sewage tanks, cable trunks, pipe trunks, pressure vessels, battery lockers, chain lockers, inert gas plant scrubber and blower spaces and the storage rooms for CO_2, halons and other media used for fire extinguishing or inerting.

10.1.2 Should there be any unexpected reduction in or loss of the means of ventilation of those spaces that are usually continuously or adequately ventilated then such spaces should also be dealt with as dangerous spaces.

10.1.3 When it is suspected that there could be a deficiency of oxygen in any space, or that toxic gases, vapours or fumes could be present, then such a space should be considered to be a dangerous space.

10.2 Precautions on Entering Dangerous Spaces

10.2.1 The following precautions should be taken as appropriate before a potentially dangerous space is entered so as to make the space safe for entry without breathing apparatus and to ensure it remains safe whilst persons are within the space.

 1 A competent person should make an assessment of the space and a responsible officer to take charge of the operation should be appointed—see 10.3

 2 The potential hazards should be identified—see 10.4

 3 The space should be prepared and secured for entry—see 10.5

 4 The atmosphere of the space should be tested—see 10.6

 5 A 'permit-to-work' system should be used—see 10.7

 6 Procedures before and during the entry should be instituted—see 10.8 and 10.9.

10.2.2 Where the procedures listed at 1 to 4 in the previous paragraph have been followed and it has been established that the atmosphere in the space is or could be unsafe then the additional requirements including the use of breathing apparatus specified in 10.10 should also be followed.

10.2.3 No one should enter any dangerous space to attempt a rescue without taking suitable precautions (see 10.10) for his own safety since not doing so would put his own life at risk and almost certainly prevent the person he intended to rescue being brought out alive.

10.3 Duties and Responsibilities of a Competent Person and of a Responsible Officer

10.3.1 A competent person is a person capable of making an informed assessment of the likelihood of a dangerous atmosphere being present or arising subsequently in the space. This person should have sufficient theoretical knowledge and practical experience of the hazards that might be met in order to be able to assess whether precautions are necessary. This assessment should include consideration of any potential hazards associated with the particular space to be entered. It should also take into consideration dangers from neighbouring or connected spaces as well as the work that has to be done within the space.

10.3.2 A responsible officer is a person appointed to take charge of every operation where entry into a potentially dangerous space is necessary. This officer may be the same as the competent person (see 10.3.1 above) or another officer. Both the competent person and/or the responsible officer may be a shore-side person.

10.3.3 It is for the responsible officer to decide on the basis of the assessment the procedures to be followed for entry into a potentially dangerous space. These will depend on whether the assessment shows:—
(a) there is no conceivable risk to the life or health of a person entering the space then or at any future time;
(b) there is no immediate risk to health and life but a risk could arise during the course of work in the space;
(c) the risk to life or health is immediate.

10.3.4 Where the assessment shows that there is no conceivable risk to health or life and that conditions in the space will not change entry may be made without restriction. Similarly an assessment could be made that there is a risk which is then entirely eliminated with no foreseeable chance whatsoever of it recurring. Entry thereafter could also be made without restriction.

10.3.5 Where the assessment shows that there is no immediate risk to health and life but that a risk could arise during the course of work in the space the precautions described in paragraphs 10.4.1 to 10.9.6 should be taken as appropriate.

10.3.6 Where the risk to life or health is immediate then the additional requirements specified in 10.10.1 to 10.10.8 are necessary.

10.3.7 For inland water vessels such as harbour craft either or both the competent person and the responsible officer may only be available from shore-based personnel. No entry into a potentially dangerous space should be made in these circumstances until such suitably qualified persons are available.

10.4 Identifying Potential Hazards

Oxygen Deficiency

10.4.1 If an empty tank or other confined space has been closed for a time the oxygen content may have been reduced due to the oxygen combining with steel in the process of rusting.

10.4.2   Lack of oxygen may occur in boilers or other pressure vessels particularly where oxygen absorbing chemicals have been used to prevent rusting.

10.4.3   Depletion of oxygen may occur in cargo spaces when oxygen absorbing cargoes, for example, oil cake and other vegetable and animal oil bearing products, certain types of wood cargoes, steel products, iron and steel swarf etc, are or have been carried.

10.4.4   Oxygen deficiency can also occur in cargo holds eg when carrying ore concentrates even though the hatch covers have been removed and the discharge of cargo has commenced.

10.4.5   After discharge of volatile cargoes sufficient vapours may remain in tanks to cause oxygen deficiency.

10.4.6   Hydrogen may occur in a cathodically-protected cargo tank used for ballast but will tend to disperse quickly when tank covers are opened. Pockets of hydrogen may, however, still exist in the upper parts of the compartment, thus displacing the oxygen (as well as creating a possible explosion hazard).

10.4.7   If carbon dioxide, steam or other fire extinguishing chemical has been discharged to extinguish or prevent a fire, the oxygen content of the space would be reduced.

10.4.8   The use of inert gas in the cargo tanks of tankers and gas carriers or in the inter-barrier spaces of gas carriers results in only minimal amounts of oxygen being present.

10.4.9   The special conditions of carriage for reactive substances may require cargo tank ullage spaces, adjacent cargo tanks, cofferdams, inter-barrier spaces and void spaces to contain inert gas.

Toxicity of Oil Cargoes

10.4.10   Hydrocarbon gases are flammable as well as toxic and may be present in fuel or cargo tanks which have contained crude oil or its products.

10.4.11   Hydrocarbon gases or vapours may also be present in pump rooms and cofferdams, duct keels or other spaces adjacent to cargo tanks due to the leakage of cargo.

10.4.12   The components in the vapour of some oil cargoes, such as benzene and hydrogen sulphide are very toxic.

Toxicity of Other Substances

10.4.13   Cargoes carried in chemical tankers or gas carriers may be toxic.

10.4.14   There is the possibility of risk of leakage from drums of chemicals or other packages of dangerous goods where there has been mishandling or incorrect stowage or damage due to heavy weather.

10.4.15   The trace components in inert gas such as carbon monoxide, sulphur dioxide, nitric oxide and nitrogen dioxide are very toxic.

10.4.16   The interaction of vegetables or animal oils or sewage with sea water may lead to the release of hydrogen sulphide which is very toxic.

10.4.17   Hydrogen sulphide or other toxic gases may be generated where the residue of grain or similar cargoes permeates into or chokes bilge pumping systems.

10.4.18   The chemical cleaning, painting or the repair of tank coatings may involve the release of solvent vapours.

Flammability

10.4.19   Flammable vapours may still be present in cargo or other tanks that have contained oil products or chemical or gas cargoes.

10.4.20   Cofferdams and other spaces that are adjacent to cargo and other tanks may contain flammable vapours should there have been leakage into the space.

Other Hazards

10.4.21   Although the inhalation of contaminated air is the most likely route through which harmful substances enter the body, some chemicals can be absorbed through the skin.

10.4.22   Some of the cargoes carried in chemical tankers and gas carriers are irritant or corrosive if permitted to come into contact with the skin.

10.4.23   The disturbance of rust, scale, or sludge residues of cargoes of animal, vegetable or mineral origin, or of water that could be covering such substances may lead to the release of toxic or flammable gases.

10.5   Preparing and Securing the Space for Entry

10.5.1   When opening the entrance to a dangerous space care should be taken to avoid the effects of a possible release of pressure or vapour from the space.

10.5.2   The space should be isolated and secured against the ingress of dangerous substances by blanking off pipe lines or other openings or by closing valves. Valves should then be tied or some other means used to indicate that they are not to be opened.

10.5.3   If necessary, the space should be cleaned or washed out to remove as far as practicable any sludge or other deposit liable to give off dangerous fumes. Special precautions (see 10.10) may be necessary when undertaking such work for the reasons given in 10.4.23 above.

10.5.4   The space should be thoroughly ventilated by either natural or mechanical means to ensure (by testing—see 10.6) that all harmful gases are removed and no pockets of oxygen deficient atmosphere remain.

10.5.5.   Compressed oxygen should not be used to ventilate any space.

10.5.6   The officers on watch, or persons in charge, on the bridge, on the deck, in the engine room, or the cargo control room should be informed, as necessary, of any space that is to be entered so that they do not, for example, stop fans, start equipment or open valves by remote control, close escape doors or pump cargo or ballast into the space and appropriate warning notices should be placed on the relevant controls or equipment. Where necessary pumping operations or cargo movements should be suspended when entry is being made into a dangerous space.

10.6 Testing the Atmosphere of the Space

10.6.1   Testing of a space should be carried out only by persons trained in the use of the equipment.

10.6.2   Testing should be carried out before entry and at regular intervals thereafter.

10.6.3   If possible, the testing of the atmosphere before entry should be made by remote means. If this is not possible, the person selected to enter the space to test the atmosphere should only do so in accordance with the additional precautions specified in 10.10, which include the wearing of breathing apparatus.

10.6.4   Where appropriate, the testing of the space should be carried out at different levels.

10.6.5   Personal monitoring equipment designed purely to provide a warning against oxygen deficiency and hydrocarbon concentrations when there is a change in conditions should not be used as a means of determining whether a dangerous space is safe to enter.

Testing for Oxygen Deficiency

10.6.6   A steady reading of at least 20% oxygen by volume on an oxygen content meter should be obtained before entry is permitted.

10.6.7   A combustible gas indicator cannot be used to detect oxygen deficiency.

Testing for Flammable Gases or Vapours

10.6.8   The combustible gas indicator (sometimes called an explosimeter) detects the amount of flammable gas or vapour in the air. An instrument capable of providing an accurate reading at low concentrations should be used to judge whether the atmosphere is safe for entry.

10.6.9   Combustible gas detectors are calibrated on a standard gas and when testing for other gases and vapours reference should be made to the calibration curves supplied with the instrument. Particular care is required should accumulations of hydrogen be suspected.

10.6.10   In deciding whether the atmosphere is safe to work in without being overcome, a 'nil' reading on a suitably sensitive combustible gas indicator is desirable but, where the readings have been steady for some time, up to 1% of lower flammable limit may be accepted, eg for hydrocarbons in conjunction with an oxygen reading of at least 20% by volume.

10.6.11   Direct measurement of trace components of inert gas (see 10.4.15) is not required when the gas freeing of the atmosphere of a cargo tank reduces the hydrocarbon concentration from about 2% by volume to 1% of lower flammable limit or less in conjunction with a steady oxygen reading of at least 20% by volume, because this is sufficient to dilute the components to a safe concentration. If, before the commencement of gas freeing, the hydrocarbon concentration of a tank containing inert gas is below 2% by volume due to excessive purging by inert gas, then additional gas freeing is necessary to remove toxic products introduced with the inert gas. It is difficult to measure the quantities of these toxic products at the safe level without specialised equipment and trained personnel. If this equipment is not available for use, the period of gas freeing should be considerably extended.

Testing for Toxic Gases

10.6.12   The presence of certain gases and vapours on chemical tankers and gas carriers is detected by fixed or portable gas or vapour detection equipment. The readings obtained by this equipment should be compared with the occupational exposure limits for the contaminant given in international industry safety guides or the latest edition of the Health and Safety Executive Guidance Note EH-40 Occupational Exposure Limits. These occupational exposure limits provide guidance on the levels of exposure to toxic substances which should not be exceeded if the health of persons is to be protected. However, it is necessary to know for which chemical a test is being made in order to use the equipment correctly and it is important to note that not all chemicals may be tested by these means.

10.6.13   When a toxic chemical is encountered for which there is no means of testing then the additional requirements specified in 10.10 should also be followed.

10.6.14   A combustible gas indicator will probably not be suitable for measuring levels of the gas at or around its occupational exposure limit, where there is solely a toxic, rather than a flammable, risk. This level will be much lower than the flammable limit, and the indicator will probably not be sufficienty sensitive to give accurate readings.

10.7   Use of a Permit-to-Work System

10.7.1   Entry into a dangerous space should be planned in advance and use should preferably be made of a 'permit-to-work' system. If, during the course of the operation, unforeseen difficulties or hazards develop, the work should be stopped and the space evacuated so that the situation can be fully assessed. Permits should be withdrawn and only issued after the situation has been re-assessed. 'Permits-to-work' should be revised as appropriate. Details of the arrangements to be followed in a 'permit-to-work' system are described in Chapter 7 which includes a specimen of a 'permit-to-work'.

10.7.2   For situations for which a well established safe system of work exists a check-list may exceptionally be accepted as an alternative to a full 'permit-to-work' provided that the principles of the 'permit-to-work' system are covered and the risks arising in the dangerous space are low.

10.7.3   On expiry of the 'permit-to-work', everyone should leave the space and the entrance to the space should be closed or otherwise secured against entry or alternatively, where the space is no longer a dangerous space, declared safe for normal entry.

10.8   Procedures and Arrangements Before Entry

10.8.1   Access to and within the space should be adequate and well illuminated.

10.8.2   No matches, welding or flame cutting equipment, electrical equipment or other sources of ignition should be taken or put into the space unless the master or responsible officer is satisfied that it is safe to do so.

10.8.3 In all cases rescue and available resuscitation equipment should be positioned ready for use at the entrance to the space. Rescue equipment means breathing apparatus together with fully charged spare cylinders of air, life lines and rescue harnesses, and torches or lamp, approved for use in a flammable atmosphere, if appropriate. A means of hoisting an incapacitated person from the confined space should also be readily available when appropriate.

10.8.4 The number of persons entering the space should be limited to those who actually need to work in the space and who could be rescued should an emergency occur.

10.8.5 At least one attendant should be detailed to remain at the entrance to the space whilst it is occupied.

10.8.6 A system of communication should be agreed and tested by all involved to ensure that any person entering the space can keep in touch with the person stationed at the entrance.

10.8.7 A system of communication should be established between the attendant at the entrance to the space and the officer on watch.

10.8.8 Before entry is permitted it should be established that entry with breathing apparatus is possible. The extent to which the use of breathing apparatus or life lines or rescue harnesses would cause any difficulty of movement within any part of the space, or would cause problems if any incapacitated person had to be removed from the space, should also be examined.

10.8.9 Lifelines of rescue harnesses should be long enough for the purpose and be easily detachable by the wearer should they become entangled, but should not otherwise come away from the rescue harnesses.

10.9 **Procedures and Arrangements During Entry**

10.9.1 Ventilation should continue during the period that the space is occupied and during temporary breaks. In the event of a failure of the ventilation system any persons in the space should leave immediately.

10.9.2 The atmosphere should be tested periodically whilst the space is occupied and persons should be instructed to leave the space should there be any deterioration in the conditions.

10.9.3 If unforeseen difficulties or hazards develop, the work in the space should be stopped and the space evacuated so that the situation can be re-assessed.

10.9.4 If a person in a space feels in any way adversely affected he should give the pre-arranged signal to the attendant standing by the entrance and immediately leave the space.

10.9.5 When available a rescue harness should be worn to facilitate recovery in the event of an accident.

10.9.6 Should an emergency occur the general (or crew) alarm should be sounded so that back-up is immediately available to the rescue team.

10.10 Additional Requirements for Entry into a Space where the Atmosphere is Suspect or Known to be Unsafe

10.10.1   If the atmosphere is considered to be suspect or unsafe to enter without breathing apparatus, then the space should only be entered if it is essential for testing purposes, the working of the ship, for the safety of life or for the safety of the ship. The number of persons entering the space should be the minimum compatible with the work to be performed.

10.10.2   Breathing apparatus should always be worn (see 10.12). Respirators cannot be used as they do not provide a supply of clean air from a source independent of the atmosphere within the space.

10.10.3   Except in the case of emergency, or where impracticable because movement in the space would be seriously impeded, two air supplies as described in 10.12.3 should be available to the wearer of the breathing apparatus who is required to work in a dangerous space. The wearer should normally use the continuous supply provided from outside the space and he should immediately make his way out of the space should it be necessary to change over to the self-contained supply.

10.10.4   During occupation of the space, precautions should be taken to safeguard the continuity of the outside source of air to the wearer of breathing apparatus. Special attention should be given to supplies originating from the engine room.

10.10.5   Where remote testing of the space (as recommended in 10.6.3 above) is not reasonably practicable, or where a brief inspection only is required, a single air supply may be acceptable provided prolonged presence in the space is not required and the wearer of the breathing apparatus is so situated that he can be hauled out immediately in case of emergency.

10.10.6   Rescue harnesses should be worn. Wherever practicable lifelines should be used. Lifelines should be attended by a person stationed at the entrance who has been trained how to pull an unconscious person from a dangerous space. Where the dangerous space to be entered requires the possible use of hoisting equipment to effect rescue, arrangements should be made to ensure that persons would be available to operate it as soon as necessary.

10.10.7   When appropriate, portable lights and other electrical equipment should be of a type approved for use in a flammable atmosphere.

10.10.8   Should there be any hazard due to chemicals, whether in liquid, gaseous or vapour form, coming into contact with the skin and/or eyes then protective clothing should be worn.

10.11   Drills and Rescue

10.11.1   Regular drills simulating the rescue of an incapacitated person from a dangerous space should be conducted to prove the feasibility of the ship's rescue plan under different and difficult circumstances. A real-weight dummy may be used for this purpose. If necessary, the space selected should be made safe for the exercise. Alternatively drill may be held, for operational convenience, in non-dangerous spaces provided that these spaces realistically simulate conditions expected in actual dangerous spaces on the ship. A drill should normally be held soon after signing on a new crew or if there has been a substantial change in crew members. Each drill should be recorded in the official log book.

10.11.2 Any attempt to rescue a person who has collapsed within a space should be based on a pre-arranged plan. Every ship will have its own individual problems each of which may require a different rescue procedure, and the plan should take into account the design of the ship and of the equipment and manpower on board. Allocation of personnel to relieve or back-up those first into the space should also be borne in mind.

10.11.3 If there are indications through the agreed system of communication or otherwise, that the person in the space is being affected by the atmosphere, the person outside the space should immediately raise the alarm. ON NO ACCOUNT SHOULD THE PERSON STATIONED AT THE ENTRANCE TO THE SPACE ATTEMPT TO ENTER IT BEFORE ADDITIONAL HELP HAS ARRIVED. NO ONE SHOULD ATTEMPT A RESCUE WITHOUT WEARING BREATHING APPARATUS AND A RESCUE HARNESS AND, WHENEVER POSSIBLE, USE OF A LIFELINE.

10.11.4 If air is being supplied through an air line to the person who is unwell, a check should be made immediately that his air supply is being maintained at the correct pressure.

10.11.5 On reaching an incapacitated person, unless he is gravely injured, eg a broken back, he should be removed from the dangerous space as quickly as possible. It is emphasised, however, that restoration of the casualty's air supply at the earliest possible moment has always to be the first priority.

10.12 Breathing Apparatus and Resuscitation Equipment

10.12.1 No one should enter a space, even to effect a rescue, where the atmosphere is unsafe or suspect without wearing breathing apparatus.

10.12.2 The two air supplies normally to be available to the wearer of a breathing apparatus working in a dangerous space (see 10.10.3) will usually comprise a continuous supply from outside the space and a self-contained supply to enable the wearer to escape to a safe atmosphere in the event of difficulty with, or failure of, the continuous supply. It should not be necessary to remove any part of the equipment or any protective clothing to change over to the self contained supply.

10.12.3 Equipment for use with two air supplies may consist of:—

(a) a conventional self-contained breathing apparatus of the open circuit compressed air type that is approved to British Standard 4667: Part 2 or equivalent standard and has been additionally tested for use with an air line connection as required by Section 3.6.3 of the British Standard or equivalent; or [BS 4667 Part 2: 1974]

(b) a compressed air line breathing apparatus incorporating an emergency self-contained supply. The compressed air line breathing apparatus should be of the demand valve type and should be approved to British Standard 4667 Part 3 or equivalent standard. The emergency self-contained supply should comply with the relevant parts of Part 3 of the British Standard as required by Section 6 of Part 4 of that standard or equivalent and the minimum cylinder capacity should be 400 litres of free air as required by section 5.10 of Part 4 of the British Standard or its equivalent. [BS 4667: Part 3: 1974; Part 4: 1989; Part 4, section 5.10.1, Note 2]

The capacity of the self-contained supply should be sufficient for the wearer of the breathing apparatus to escape to a safe atmosphere. When

Part 2: 1974
and Part 4: 1989

determining this capacity it should be recognised that if under stress or in difficult conditions the wearer's breathing rate may be in excess of the nominal breathing rate of 40 litres per minute referred to in Parts 2 and 4 of the British Standard.

10.12.4 Where the supply of air comes from outside the space, the responsible officer should make sure that it is continuous and is available only to those working in the space. Pipeline or hoses supplying air should be placed so that they are not likely to be so distorted that supply might be interrupted or damaged. If the purpose for which such air lines are used is not immediately apparent to persons not engaged in the entry, then notices should be posted at appropriate positions. When a mechanical driven pump is being used it should frequently be checked carefully to ensure that it continues to operate properly. Any air pumped directly into a pipeline or put into reserve bottles requires to be filtered and should be as fresh as possible. Pipelines or hoses used to supply air should be thoroughly blown through to remove moisture and freshen the air before connection to breathing apparatus and face masks. It is essential that where the air supply is from a compressor sited in a machinery space, the engineer of the watch be informed so that the compressor is not shut-down until the work is completed.

10.12.5 Everyone likely to use breathing apparatus should be instructed by a competent person in its proper use.

10.12.6 The master, or responsible officer, and the person about to enter the space should undertake the full pre-wearing check and donning procedures recommended in the manufacturer's instructions. In particular they should check:—

1 that there will be sufficient clean air at the correct pressure;

2 that low pressure alarms are working properly;

3 that the facemask fits correctly against the user's face so that, combined with pressure of the air coming into the mask, there will not be an ingress of oxygen deficient air or toxic vapours when the user inhales. It should be noted that facial hair or spectacles may prevent the formation of an air-tight seal between a person's face and the facemask;

4 that the wearer of the breathing apparatus understands whether or not his air supply may be shared with another person and if so is also aware that such procedures should only be used in an extreme emergency;

5 that when work is being undertaken in the space the wearer should keep the self-contained supply for use when there is a failure of the continuous supply from outside the space.

10.12.7 When in a dangerous space:—

1 No one should remove his own breathing apparatus.

2 Breathing apparatus should not be removed from a person unless it is necessary to save his life.

10.12.8 It is recommended that resuscitators of an appropriate kind should be provided where any person may be required to enter a dangerous space. Where entry is expected to occur at sea the ship should be provided with appropriate equipment. Otherwise entry should be deferred until the ship has docked and use can be made of shore side equipment.

10.13 Maintenance of Equipment

10.13.1   All breathing apparatus, rescue harnesses, lifelines, resuscitation equipment and any other equipment provided for use in, or in connection with, entry into dangerous spaces, or for use in emergencies, should be properly maintained, inspected periodically and checked for correct operation by a competent person and a record of the inspections and checks should be kept. All items of breathing apparatus should be inspected and checked for correct operation before and after use.

10.13.2   Equipment for testing the atmosphere of dangerous spaces, including oxygen meters, should be kept in good working order and, where applicable, regularly serviced and calibrated. Due regard should be paid to manufacturer's recommendations which should always be kept with the equipment.

10.14 Training, Instruction and Information

10.14.1   Employers should provide any necessary training, instruction and information to employees in order to ensure that the requirements of the Entry into Dangerous Spaces Regulations are complied with. This should include:—

1. recognition of the circumstances and activities likely to lead to the presence of a dangerous atmosphere,

2. the hazards associated with entry into dangerous spaces, and the precautions to be taken,

3. the use and maintenance of equipment and clothing required for entry into dangerous spaces,

4. instruction and drills in rescue from dangerous spaces.

10.15 Statutory Regulations

MS (Entry into Dangerous Spaces) Regs SI 1988 No. 1638

10.15.1   The Merchant Shipping (Entry into Dangerous Spaces) Regulations 1988 place obligations on the Master of a ship and his employer to ensure that procedures for safe entry and working in dangerous spaces are clearly laid down and observed on board the ship, and on persons entering or remaining in a dangerous space to do so only in accordance with these procedures. In carrying out these duties full account must be taken of the principles and guidance described in this Chapter.

CHAPTER 11

Manual lifting and carrying

11.1 Guidance to Employers

11.1.1 Many people have sustained serious back and other injuries during manual lifting or carrying operations as a result of accidents, poor organisation or unsatisfactory working methods and employers should always aim to find safer practicable alternatives to such operations on board ship.

11.1.2 Before employees are instructed to lift or carry by hand the employer should have ensured that the attendant risks to health and safety have been evaluated and due account taken of them in the training provided and the working methods used.

11.1.3 When assessing the risks and considering adequate protection full account should be taken not only of the characteristics of the load and the physical effort required but also of the working environment (ship movement, confined space, high or low temperature, physical obstacles such as steps or gangways, etc) and any other relevant factors (eg the age and health of the person, the frequency and duration of the work, etc).

11.2 Guidance to Seafarers

11.2.1 In manual lifting and carrying, the proper procedure to be followed as a matter of habit is to size up the load to be lifted, look for sharp edges, protruding nails or splinters, for greasy or other surfaces which may affect grip and for any other features which may prove awkward or dangerous; for example sacks of bulk commodities may be difficult to get off the deck.

11.2.2 The deck or area over which the load is to be moved should be free from obstructions and not slippery.

11.2.3 A firm and balanced stance should be taken close to the load with feet a little apart, not too wide, so that the lift will be as straight as possible.

11.2.4 A crouching position should be adopted, knees bent and back straight to ensure that the legs do the work—keeping chin tucked in.

11.2.5 The load should be gripped with the whole of the hand—not fingers only. If there is insufficient room under a heavy load to do this a piece of wood should be put underneath first.

11.2.6 The size and shape of the load are not good guides to its weight or weight distribution. If this information is not available a careful trial lift should be made, and if there is any doubt whether the load can be managed by one man help should be provided.

11.2.7 When two or more men are handling a load, it is preferable that they should be of similar height. The actions of lifting, lowering and carrying should, as far as possible, be carried out in unison to prevent strain and any tendency for either person to overbalance.

11.2.8 The load should be lifted by straightening the legs, keeping it close to the body. The body should not be twisted as this will impose undue strain.

11.2.9 If the lift is to a high level, it may be necessary to do it in two stages; first raising the load on to a bench or other support and then completing the lift to the full height, with a fresh grip.

11.2.10 The procedure for putting a load down is the reverse of that for lifting, the legs should do the work of lowering—knees bent, back straight and the load close to the body. Care should be taken not to trap fingers. The load should not be put down in a position where it is unstable.

11.2.11 A load should always be carried in such a way that it does not obscure vision, so that any obstruction in the passageway can be seen.

11.2.12 Suitable shoes or boots should be worn for the job. Protective toecaps help to guard toes from crushing if the load slips; they can sometimes also be useful when putting the load down to take the weight while hands are removed from underneath.

11.2.13 Clothing should be worn which does not catch in the load and which gives some body protection.

11.2.14 Where the work is very strenuous, for example because of the weight of the load, repetitive efforts over a period or environmental factors such as confined space or extremes of temperature, rest should be taken at suitable intervals, to allow muscles, heart and lungs to recover; fatigue makes accidents more likely on work of this kind.

11.2.15 Whenever possible, manual lifting and carrying should be organised in such a way that each man has some control over his own rate of work.

CHAPTER 12

Tools and materials

12.1 Work equipment and the employer

12.1.1 The employer should ensure that any machine, tool, installation or other work equipment made available for the use of seafarers is suitable for the work in hand and for the conditions in which that work is to be carried out, and that it is so maintained that it may be used without impairing health or safety.

12.1.2 If it is not possible to remove a health or safety hazard, appropriate measures should be taken to ensure that the risk is minimised. For example, the maintenance and use of the equipment should, where appropriate, be restricted to designated persons. Adequate information and training in the safe use of equipment (including, where appropriate, comprehensible written instructions) should be provided for all users.

12.2 Hand tools

12.2.1 A tool is designed for one particular function and no other. It should be treated with respect. The material of which it is made is appropriate to the intended purpose but usually not to others. Files are hard but brittle; screwdriver shanks bend where levers do not, and pliers may slip on nuts.

12.2.2 For every job, the proper tools in the right sizes should be available and used. Tools used for a purpose for which they were not designed may cause injury to the user and damage to the workpiece and the tools.

12.2.3 Damaged or worn tools should not be used. Handles of hammers, screwdrivers and chisels should be secure; wooden handles should be straight-grained, smooth and without splinters. Punches and cold chisels with jagged heads should not be used. Cutting edges should be kept sharp and clean. Faces of hammers, punches and spanners should be true. Repair and dressing of the tools should be carried out by a competent person.

12.2.4 When not in use, they should be stowed tidily in a suitable tool rack, box or carrier, with cutting edges protected.

12.2.5 Wherever practicable, a tool in use should be directed away from the body to avoid injury should the tool or workpiece slip.

12.2.6 Both hands should be kept behind the cutting edge of a wood chisel.

12.2.7 A cold chisel is best held between thumb and base of index finger with thumb and fingers straight, palm of hand facing towards the hammer blow.

12.2.8 A saw should not be forced, it should be pushed with a light, even movement.

12.3 Portable electric, pneumatic and hydraulic tools and appliances

12.3.1 Power-operated tools may be dangerous unless properly maintained, handled and used.

12.3.2 The risk of electric shock is greatly increased either by perspiration or in locations which are damp, humid or have large conductive (metal) surfaces. In such conditions power tools should be operated from extra low voltage supplies (not more than 50 volts AC with a maximum of 30 V to earth or 50 V DC). However even 50 V can be lethal in particularly severe conditions, when lower voltages should be used.

Where it is not practicable to use low voltages then other precautions such as a local isolating transformer supplying one appliance only or a high sensitivity earth leakage circuit breaker (also known as a residual current device) should be used. The risk associated with the use of portable electric tools also applies to portable electric hand lamps. The supply to these appliances should not exceed 24 volts. (See also 9.3.9).

12.3.3 Double-insulated tools (where the exposed metal parts are not designed for earth connection) are not recommended for use on ships because water (which may be salt-laden) can provide a contact between live parts and the casing, increasing the risk of a fatal shock when the tool is used.

12.3.4 The power supply lead and connections should be inspected before a tool is used; defects should be repaired and the tool tested by a competent person before its re-use.

BS 6500: 1990

12.3.5 The flexible cables of electric tools should comply with the relevant British Standard and be provided with proper connections to the power supply. The tools should be properly earthed.

12.3.6 The fuse or circuit-breaker on the line of supply to electric power tools should be of the minimum rating practicable. This is most important if double-insulated tools are used.

12.3.7 Electric leads and the hoses of pneumatic and hydraulic tools should be kept clear of damage from nails, sharp edges, hot surfaces, oil and chemicals etc. Where leads or hoses pass through doorways or other openings, the doors etc should be secured open. Where they trail across decks or passageways, leads or hoses should wherever possible be suspended high enough to give clearance over men passing.

12.3.8 Whip-lash from pneumatic hoses in the event of breakage of couplings, may be prevented by fitting chain linkages between sections of the hose or alternatively incorporating safety valves which close off the lines.

12.3.9 Accessories or tool pieces should be absolutely secure in the tool. In particular, retaining springs, clamps, locking levers and other built-in safety devices on pneumatic tools should be replaced after the toolpiece (drill, bit, chisel etc) is changed. Serious injuries can be caused if any of these are omitted, since the toolpiece may be ejected with considerable force when power is applied.

12.3.10 Accessories or fitments should not be fixed or replaced while the tool is connected to a source of power.

12.3.11 Where a safety guard is needed for a particular operation, it should be securely fixed before work begins; if it is removed for changing an accessory, it should be replaced immediately.

12.3.12 During a temporary interruption of work, power tools should be switched off and disconnected from the source of power and left in a safe position with leads clear of passageways. A check that the switch or control is off should always be made before the tool is reconnected.

12.3.13 Where the work operation causes high noise levels, hearing protection should be worn. Where flying particles may be produced, the face and eyes should be protected (see Chapter 5).

12.3.14 The vibration caused by reciprocating tools (pneumatic drills, hammers, chisels etc) or high-speed rotating tools (eg drills) can give rise to a permanent disablement of the hands known as 'dead' or 'white' fingers. In its initial stages, this appears as a numbness of the fingers and an increasing sensitivity to cold but in more advanced stages, the hands become blue and the finger-tips swollen. Those prone to the disability should not use such portable power tools. Others should be advised not to use them continually for more than a maximum of 30 minutes without a break.

12.4 Workshop and bench machines (fixed installations)

12.4.1 No one should operate a machine unless authorised to do so. The operator should be competent in its use and familiar with its controls. He should not attempt to use it if he has bandaged hands. (See Section 1.3 concerning garments, long hair etc).

MS (Guarding of Machinery and Safety of Electrical Equipment) Regs SI 1988 No 1636

12.4.2 All dangerous parts of machines should be provided with efficient guards which should be properly secured before the machine is put into operation. Self-adjusting guards are preferable where the position of the guard has to relate to the workpiece. Grinding machines should be fitted with eye screens which need to be renewed from time to time.

12.4.3 Guards should be made preferably in solid material. Where they are of perforated metal, mesh or bars, the openings should not be large enough to allow a finger to be inserted to reach a dangerous part.

12.4.4 Controls of machines and switches for supplementary lighting, where this is provided, should not be so placed that the operator has to lean over the machine to reach them.

12.4.5 A machine should be checked every time before use. It should not be operated when a guard or safety device is missing, incorrectly adjusted or defective or when it is itself in any way faulty.

12.4.6 If defective in any respect, the machine should be isolated from its source of power pending adjustment or repair. Only a competent person should attempt repairs. Unskilled interference with electrical equipment in particular is highly dangerous.

12.4.7 Work benches should be well lit and some machines may required individual supplementary lights.

12.4.8   Working areas should be kept uncluttered and, as far as practicable, free of litter and spilled oil. Loose gear, tools and equipment not required for immediate use should be cleared away and properly stowed.

12.4.9   Swarf (metal turnings, filings and the like) should not be allowed to pile up round a machine. The machine should be stopped for its removal. A rake or similar device should be used for the purpose, never the bare hand.

12.4.10   A heavy item of equipment brought into a workshop for repair should be made secure against accidental movement.

12.4.11   Appropriate eye and face protection should be worn during chipping, scaling, wirebrushing, grinding and similar work where particles may fly; this is a special risk in turning brass (see Chapter 5).

12.4.12   Where sanding or other processes generate a lot of dust in the air, dust masks or respirators should be worn (see Chapter 5).

12.4.13   Other people working in the area may need the protection indicated in either of the two preceding paragraphs.

12.4.14   Before a lathe or drill is started, the chuck key should be removed and the operator should make sure that other people are clear of the machine.

12.4.15   A machine should be stopped when not in use, even if it is expected to be left unattended for a few moments only. The machine should be rechecked on every occasion before being started up again in case controls, guards etc have been altered or moved while the machine has been left unattended.

12.4.16   Where a machine is driven by a V-belt in conjunction with a stepped pulley, and alterations in spindle speed require a change in the belt position, means should be provided if practicable for the belt tension to be eased during that operation; the position of the belt should never be changed while the machine is running.

12.4.17   Work pieces for drilling or milling should be at all times securely held by a machine vice or clamp.

12.4.18   Material projecting beyond the headstock of a lathe should be securely fenced.

12.5   Abrasive wheels

12.5.1   Abrasive wheels should be selected, mounted and used only by competent persons and in accordance with manufacturers' instructions.

12.5.2   Abrasive wheels are relatively fragile and should be stored and handled with care.

12.5.3   Manufacturers' instructions should be followed on the selection of the correct type of wheel for the job in hand. Generally, soft wheels are more suitable for hard material and hard wheels for soft material.

12.5.4 Before a wheel is mounted, it should be brushed clean and closely inspected to ensure that it has not been damaged in storage or transit. The soundness of a vitrified wheel can be further checked by suspending it vertically and tapping it gently. If the wheel sounds dead it is probably cracked, and should not be used.

12.5.5 A wheel should not be mounted on a machine for which it is unsuitable.

12.5.6 The wheel should fit freely but not loosely on the spindle; if the fit is unduly tight, the wheel may crack as the heat of operation causes the spindle to expand.

12.5.7 The clamping nut should be tightened only sufficiently to hold the wheel firmly. When the flanges are clamped by a series of screws, the screws should be first screwed home with the fingers and diametrically opposite pairs tightened in sequence.

12.5.8 The speed of the spindle should not exceed the stated maximum permissible speed for the wheel.

12.5.9 A strong guard should be provided and kept in position at every abrasive wheel (unless the nature of the work absolutely precludes its use) both to contain wheel parts in the event of a burst and to prevent an operator having contact with the wheel.

12.5.10 The guard should enclose as much of the wheel as possible.

12.5.11 Where a workrest is provided, it should be properly secured to the machine and should be adjusted as close as practicable to the wheel, the gap normally being not more than 1.5 mm ($^1/_{16}$ inch).

12.5.12 The side of a wheel should not be used for grinding: it is particularly dangerous when the wheel is appreciably worn.

12.5.13 The workpiece should never be held in a cloth or pliers.

12.5.14 When dry grinding operations are being carried on or when an abrasive wheel is being trued or dressed, suitable transparent screens should be fitted in front of the exposed part of the wheel or operators should wear properly fitting eye protectors.

12.6 Spirit lamps

12.6.1 Care should be taken in filling spirit lamps. Fuel should not be added to a lamp which has been in use until it has completely cooled down.

12.7 Compressed air

12.7.1 When compressed air is used, the pressure should be kept no higher than is necessary to undertake the work satisfactorily.

12.7.2 Compressed air should not be used to clean the working space.

12.7.3 In no circumstances should compressed air be directed at any part of a person's body.

12.8 Compressed gas cylinders

12.8.1   Compressed gas cylinders should always be handled with care, whether full or empty. They should be properly secured and kept upright. The arrangements for securing the cylinders should be capable of quick and easy release so that they may be readily removed in, say, the case of fire. If available, cylinder trolleys should be used to transport cylinders from one place to another.

12.8.2   The protective caps over the valve should be screwed in place when the cylinders are not in use or are being moved. Valves should be closed when the cylinder is empty.

12.8.3   Where two or more cylinders of either oxygen or a fuel gas (such as acetylene) are carried the oxygen and the fuel gas should be stowed in separate, well ventilated compartments that are not subject to extremes of temperature. The space in which acetylene or other fuel gas cylinders are stowed should have no electrical fittings or other sources of ignition and prominent and permanent 'NO SMOKING' signs should be displayed at the entrance and within the space. Empty cylinders should be segregated from full ones and so marked.

12.8.4   Special precautions as follows need to be taken in the case of cylinders of oxygen and acetylene or other fuel gases:
(a) cylinder valves, controls and associated fittings should be kept free from oil, grease and paint. Controls should not be operated with oily hands;
(b) gas should not be taken from such cylinders unless the correct pressure reducing regulator has been attached to the cylinder outlet valve;
(c) cylinders found to have leaks that cannot be stopped by closing the outlet valve should be taken to the open deck away from any sources of heat or ignition and slowly discharged to the atmosphere.

12.8.5   Identifying markings on cylinders are set out in Section 6.5.

12.9 Chemical agents

12.9.1   A chemical from an unlabelled container should never be used unless its identity has been positively established.

12.9.2   Chemicals should always be handled with the utmost care. Eyes and skin should be protected from accidental exposure or contact.

12.9.3   Manufacturers' or suppliers' advice on the correct use of the chemicals should always be followed. Some cleaning agents, even though used domestically, for example, caustic soda and bleaches, may burn the skin.

12.9.4   Chemicals should not be mixed unless it is known that dangerous reactions will not be caused.

CHAPTER 13

Welding and flamecutting operations

13.1 General

13.1.1 Welding and flamecutting elsewhere than in the workshop should generally be the subject of a 'permit-to-work' (see Chapter 7).

13.1.2 Operators should be competent in the process, familiar with the equipment to be used and instructed where special precautions need to be taken.

13.1.3 Where portable lights are needed to provide adequate illumination, they should be clamped or otherwise secured in position, not hand-held, with leads kept clear of the working area.

13.1.4 Harmful fumes can be produced during these operations especially from galvanising, paint, etc. Oxygen in the atmosphere can be depleted when using gas cutting equipment and noxious gases may be produced when welding or cutting. Special care should therefore be taken when welding and flamecutting in enclosed spaces to provide adequate ventilation. The effectiveness of the ventilation should be checked at intervals while the work is in progress. In confined spaces, breathing apparatus may be required.

13.1.5 Welding and flamecutting equipment should be inspected before use by a competent person to ensure that it is in a serviceable condition. All repairs should be carried out by a competent person.

13.2 Protective clothing

BS 2653: 1955

13.2.1 Protective clothing and equipment complying with the relevant British Standard Specifications should be worn by the operator and as appropriate by those working with him to protect them from particles of hot metal and slag and from accidental burns and their eyes and skin from ultra-violet and heat radiation.

13.2.2 The operator should normally wear:
(a) welding helmet with suitably coloured transparent eye piece. Eye goggles or a hand-held shield may be suitable alternatives in appropriate circumstances;
(b) leather working gloves;
(c) leather apron (in appropriate circumstances);
(d) long-sleeved natural fibre boiler suit or other approved protective clothing.

13.2.3 Clothing should be free of grease and oil and other flammable substances.

13.3 Precautions against fire and explosion

13.3.1   Before welding, flamecutting or other hot work is begun, a check should be made that there are no combustible solids, liquids or gases, at, below or adjacent to the area of the work, which might be ignited by heat or sparks from the work.

13.3.2   Welding or other hot work should never be undertaken on surfaces covered with grease, oil or other flammable or combustible substances.

13.3.3   When welding is to be done in the vicinity of open hatches, suitable screens should be erected to prevent sparks dropping down hatchways or hold ventilators. Where necessary, combustible materials and dunnage should be moved to a safe distance before commencing operations.

13.3.4   Port holes and other openings through which sparks may fall should be closed where practicable.

13.3.5   Where work is being done close to or at bulkheads, decks or deckheads, the remote sides of the divisions should be checked for materials and substances which may ignite, and for cables, pipelines or other services which may be affected by the heat.

13.3.6   Cargo tanks, fuel tanks, cargo holds or other tanks or spaces that have contained flammable substances should be certified as being free of flammable gases before any repair work is commenced. The testing should include, as appropriate, the testing of adjacent spaces, double bottoms, cofferdams, etc. Further tests should be carried out at regular intervals and before hot work is recommenced following any suspension of the work. When preparing tankers or similar ships all tanks, cargo pumps and pipelines should be thoroughly cleaned and particular care taken with the draining and cleaning of pipelines that cannot be directly flushed using the ship pumps.

13.3.7   Welding and flamecutting operations should be properly supervised and kept under regular observation. Suitable fire extinguishers should be kept at hand ready for use during the operation. A person with a suitable extinguisher should also be stationed to keep watch on areas not visible to the welder which may be affected.

13.3.8   In view of the risk of delayed fires resulting from the use of burning or welding apparatus, appropriate frequent checks should be made for at least two hours after cessation of the work.

13.4 Electric welding equipment

13.4.1   In order to minimise risk from electric shock, electric welding power sources for shipboard use should have a direct current (DC) output not exceeding 70 V, with a minimum ripple. Further information on DC power sources is given in 13.4.10.

13.4.2   When DC equipment is not available, then AC output power sources may be used providing they have an integral voltage limiting device to ensure that the idling voltage (the voltage between electrode and work piece before an arc is struck between them) does not exceed 25 V rms. The proper function of the device (which may be affected by dust or humidity) should be checked each time a welding set is used. Some voltage limiting devices are affected by their angle of tilt from the vertical, so it is important that they are mounted and used in the position specified by the manufacturers. This requirement can be affected by adverse sea conditions.

13.4.3 A 'go and return' system utilising two cables from the welding set should be adopted; the welding return cable should be firmly clamped to the workpiece.

13.4.4 The 'return' cable of the welding set and the workpiece or workpieces should be separately earthed to the ship's structure. The use of a single cable with hull return is not recommended.

13.4.5 To avoid voltage drop in transmission, the lead and return cables should be of the minimum length practicable for the job and of an appropriate cross-section.

13.4.6 Cables should be inspected before use; if the insulation is impaired or conductivity is reduced, they should not be used.

13.4.7 Cable connectors should be fully insulated when connected, and so designed and installed that current carrying parts are adequately recessed when disconnected.

13.4.8 Electrode holders should be fully insulated so that no live part of the holder is exposed to touch, and, where practicable, should be fitted with guards to prevent accidental contact with live electrodes and as protection from sparks and splashes of weld metal.

13.4.9 A local switching arrangement or other suitable means should be provided for rapidly cutting off current from the electrode should the operator get into difficulties and also for isolating the holder when electrodes are changed.

13.4.10 The direct current output from power sources should not exceed 70 volts open circuit. The ripple on the output from the power source should not exceed the values of the table below. The ripple magnitudes are expressed as percentages of the DC, and the ripple peak is that with the same polarity as the DC.

Ripple Frequency, Hz	50/60	300	1200	2400
Max. RMS O/C voltage ripple, (%)	5	6	8	10
Max. peak O/C voltage ripple, (%)	10	12	16	20

13.4.11 The conditions in the table in 13.4.10 are normally met by DC generators incorporating commutators and by rectifier power sources having a 3 phase bridge rectifier operating from a 3 phase 50/60 Hz supply. Rectifier power sources should not be operated from a power supply of less than 50 Hz.

13.4.12 Should it be necessary to use a power source with a DC output having a ripple magnitude in excess of those stated in the table, for example a single phase rectifier power source, then a voltage limiting device should be incorporated in the power source to ensure that the idling voltage does not exceed 42 V.

13.5 Precautions to be taken during electric-arc welding

13.5.1 The welding operator should wear the protective clothing specified in 13.2.2 but should additionally wear non-conducting safety footwear. Clothing should be kept as dry as possible as some protection against electric shock; it is particularly important that gloves should be dry because wet leather is a good conductor.

13.5.2 An assistant should be in continuous attendance during welding operations. He should be alert to the risk of accidental shock to the welder, ready to cut off power instantly, raise the alarm and apply artificial respiration without delay. The desirability of a second assistant should be considered if the work is to be carried out in difficult conditions.

13.5.3 Where persons other than the operator are likely to be exposed to harmful radiation or sparks from electric arc welding, they should be protected by screens or other effective means.

13.5.4 In restricted spaces, where the operator may be in close contact with the ship's structure or is likely to make contact in the course of ordinary movements, protection should be provided by dry insulating mats or boards.

13.5.5 There are increased risks of electric shock to the operator if welding is done in hot or humid conditions; body sweat and damp clothing greatly reduce body resistance. Under such conditions, the operation should be deferred until such time as an adequate level of safety can be achieved.

13.5.6 In no circumstances should a welder work while standing in water or with any part of his body immersed.

13.5.7 The electrode holder should be isolated from the current supply before a used electrode is removed and before a new electrode is inserted. This precaution is necessary because some electrode coatings have extremely low resistance. Even a flux coating which is normally insulating can become damp from sweating hands and thus potentially dangerous.

13.5.8 When the welding operation is completed or temporarily suspended, the electrode should be removed from the holder.

13.5.9 Hot electrode ends should be ejected into a suitable container; they should not be handled with bare hands.

13.5.10 Spare electrodes should be kept dry in their container until required for use.

13.6 Gas Welding and Cutting

13.6.1 Advice on the storage and handling of gas cylinders is given in Section 12.8.

13.6.2 The pressure of oxygen used for welding should always be high enough to prevent acetylene flowing back into the oxygen line.

13.6.3 Acetylene should not be used for welding at a pressure exceeding 1 atmosphere gauge as it is liable to explode, even in the absence of air, when under excessive pressure.

13.6.4 Back pressure valves should be fitted adjacent to the torch in the oxygen and acetylene supply lines.

13.6.5 Flame arrestors should be provided in the oxygen and acetylene supply lines and will usually be fitted at the low pressure side of regulators although they may be duplicated at the torch.

13.6.6 Should a backfire occur the valves on the oxygen and acetylene cylinders should be closed as soon as possible. A watch should be kept on the acetylene cylinders and should one become hot it should be immediately removed to the open, kept cool either by immersion or with copious amounts of water and the cylinder stop valve opened fully. If this cannot be done with safety, consideration should be given to jettisoning the cylinder overboard. Any acetylene cylinder suspected of overheating should be treated with care because an impact could set off an internal ignition which might cause an explosion.

13.6.7 Only acetylene cylinders of approximately equal pressures should be coupled.

13.6.8 In fixed installations, manifolds should be clearly marked with the gas they contain.

13.6.9 Manifold hose connections including inlet and outlet connections should be such that the hose cannot be interchanged between fuel gases and oxygen manifolds and headers.

13.6.10 Only those hoses specially designed for welding and cutting operations should be used to connect an oxy-acetylene blowpipe to gas outlets.

13.6.11 Any length of hose in which a flashback has occurred should be discarded.

13.6.12 The connections between hose and blowpipe, and between hoses, should be securely fixed with metal fittings such as hose bands.

13.6.13 Hoses should be so arranged that they are not likely to become kinked or tangled or be tripped over, cut or otherwise damaged by moving objects or falling metal slag, sparks etc; a sudden jerk or pull on the hose is liable to pull the blowpipe out of the operator's hands or cause a cylinder to fall or a hose connection to fail. Hoses in passageways should be covered to avoid them becoming a tripping hazard.

13.6.14 Soapy water only should be used for testing leaks in hoses.

13.6.15 Blowpipes should be lit with a special friction igniter, stationary pilot flame or other safe means.

13.6.16 Should a blowpipe-tip opening become clogged, it should be cleaned only with the tools especially designed for that purpose.

13.6.17 When a blowpipe is to be changed the gases should be shut off at the pressure-reducing regulators.

13.6.18 During a temporary stoppage or after completion of the work, supply valves on gas cylinders and gas mains should be securely closed and blowpipes, hoses and movable pipes should be removed to lockers that open on to the open deck, to prevent a build-up of dangerous concentrations of gas or fumes.

13.6.19 Oxygen should never be used to ventilate, cool or blow dust off clothing (see also Section 12.7).

CHAPTER 14

Painting

14.1 General

14.1.1 Paints may contain toxic or irritant substances, and the solvents may give rise to flammable and potentially explosive vapours, which may also be toxic; men using such paints should be warned of the particular risks arising from their use. Paints containing organic pesticides can be particularly dangerous. If the manufacturer's instructions are not given on the container, it should be ascertained at the time of supply whether any special hazards may arise from the use of the paint and also whether special methods of application should be followed. Such advice should be readily available at the time of use but the following precautions should always be taken in any case.

14.1.2 Painted surfaces should not be rubbed down dry unless it is known that the old paint is free from lead or other substance, the dust from which could be toxic if inhaled. Dust masks should be worn as protection against other dusts.

14.1.3 Rust removers are acids and contact with the skin should be avoided. Eye protection should be worn against splashes. If painting aloft or otherwise near ropes, care should be taken to avoid splashes on ropes, safety harness, lines, etc. Reference should be made to Section 15.4 on the effect of such contamination on ropes.

14.1.4 Interior and enclosed spaces should be well ventilated, both while painting is in progress and until the paint has dried.

14.1.5 There should be no smoking or use of naked lights in interior spaces during painting or until the paint has dried hard. Some vapours even in low concentrations may decompose into more harmful substances when passing through burning tobacco.

14.1.6 When painting is done in the vicinity of machinery, especially in the engine room or, for instance, from an overhead crane gantry, care should be taken to ensure that the power supply is isolated and the machinery immobilised in such a way that it cannot be moved or started up inadvertently. Appropriate warning notices should be posted (see 22.1.11). Close-fitting clothing should be worn (see Section 1.3).

14.2 Spraying

14.2.1 There are many different types of paint spraying equipment in use and operators should acquaint themselves with the manufacturer's instructions on the correct usage of the equipment prior to the operation, and how to avoid risks in use.

14.2.2 Airless spray painting equipment is particularly hazardous since the paint is ejected at very high pressure and can penetrate the skin or cause serious eye injuries to the operator or anyone else at close range. Great care should therefore be taken when this system is used.

14.2.3 Suitable protective clothing such as a combination suit, gloves, cloth hood, and eye protection should be worn during spraying.

14.2.4 Paints containing lead, mercury or similarly toxic compounds should not be sprayed in interiors.

14.2.5 A suitable respirator should be worn according to the nature of the paint being sprayed. In exceptional circumstances it may be necessary to use breathing apparatus (see Section 5.5).

14.2.6 If a spray nozzle clogs, the trigger of the gun should be locked in a closed position before any attempt is made to clear the blockage.

14.2.7 Before a blocked spray nozzle is removed or any other dismantling attempted, pressure should be relieved from the system.

14.2.8 A gun having a reversible nozzle calls for special care to ensure that hands are kept clear of the nozzle mouth when blowing through it to remove a blockage.

14.2.9 The pressure in the system should not exceed the recommended working pressure of the hose. The system should be regularly inspected for defects.

14.2.10 As an additional precaution against the hazards of a hose bursting, a loose sleeve, for example a length of 2 to 3 metres (6 to 10 feet) of old air hose, may be slipped over that portion of the line adjacent to gun and paint container.

14.3 Painting aloft, overside and from punts

14.3.1 The same precautions should be taken in painting aloft as for other work aloft (see Chapter 15).

14.3.2 Painting punts should be stable and provided with suitable fencing. Unsecured trestles and planks should not be used to give additional height.

14.3.3 The person in charge should have due regard to the strength of tides and other hazards, such as wash from passing vessels, before a painting punt is put to use.

14.3.4 A man painting overside should wear a life-line and buoyancy garment and should be under observation by a seaman on deck; a lifebuoy with a sufficient length of line attached should be ready for immediate use.

14.3.5 When painting is to be done at or near the stern or other propeller aperture, the person in charge should inform the duty engineer and deck officers so that warning notices are put up in the engine room, at the controls and on the bridge.

14.3.6 The duty engineer and deck officers should also be informed by the person in charge when seamen are painting below ship's side discharges so that they are not used until the work is completed. Notices to this effect should be attached to the relevant control valves and not taken off until the men are reported clear.

CHAPTER 15

Working aloft and outboard

15.1 General

15.1.1 A man working at a height may not be able to give his full attention to the job and at the same time guard himself against falling. Proper precautions should therefore always be taken to ensure personal safety when work has to be done aloft or when working outboard. It must be remembered that the movement of a ship in a seaway will add to the hazards involved in work of this type. A stage or ladder should always be utilised when work is to be done beyond normal reach.

15.1.2 Seamen under 18 years of age or with less than 12 months experience at sea, should not work aloft unless accompanied by an experienced seaman or otherwise adequately supervised.

15.1.3 A safety harness with lifeline or other arresting device should be continuously worn when working aloft, outboard or overside (see section 5.7). A safety net should be rigged where necessary and appropriate. Additionally, where work is done overside, buoyancy garments should be worn and a lifebuoy with sufficient line attached should be kept ready for immediate use.

15.1.4 Men should not work overside while the vessel is underway.

15.1.5 Before work is commenced near the ship's whistle, the officer responsible for the job should ensure that power is shut off and warning notices posted on the bridge and in the machinery spaces.

15.1.6 Before work is commenced on the funnel, the officer responsible should inform the duty engineer to ensure that steps are taken to reduce as far as practicable the emission of steam, harmful gases and fumes.

15.1.7 Before work is commenced in the vicinity of radio aerials, the officer responsible should inform the radio officer so that no transmissions are made whilst there is risk to the seafarer. A warning notice should be put up in the radio room.

15.1.8 Where work is to be done near the radar scanner, the officer responsible should inform the officer on watch so that the radar and scanner are isolated. A warning notice should be put on the set until the necessary work has been completed.

15.1.9 On completion of the work of the type described above, the officer responsible should, where necessary, inform the appropriate officer that the precautions taken are no longer required and that warning notices can be removed.

15.1.10 Unless it is essential, work should not be done aloft on a stage or bosun's chair in the vicinity of cargo working.

15.1.11   Care must also be taken while work is being done aloft or at a height, to avoid risks to anyone working or moving below. Suitable warning notices should be displayed. Tools and stores should be sent up and lowered by line in suitable containers which should be secured in place for stowage of tools or materials not presently being used.

15.1.12   No one should place tools where they can be accidentally knocked down and may fall on someone below, nor should tools be carried in pockets from which they may easily fall. When working aloft it is often best to wear a belt designed to hold essential tools securely in loops.

15.1.13   Tools should be handled with extra care when hands are cold or greasy and where the tools themselves are greasy.

15.2 Cradles and stages

15.2.1   Cradles should be at least 430 mm (17 inches) wide and fitted with guard rails or stanchions with taut ropes to a height of 1 metre (39 inches) from the floor. Toeboards add safety.

15.2.2   Planks and materials used for the construction of ordinary plank stages must be carefully examined to ensure adequate strength and freedom from defect.

15.2.3   Wooden components of staging should be stowed in a dry, ventilated space and not subjected to heat.

15.2.4   Ancillary equipment, lizards, blocks and gantlines should be thoroughly examined before use. A defective item should not be used.

15.2.5   When a stage is rigged overside, the two gantlines used in its rigging should at least be long enough to trail into the water to provide additional lifelines should the operator fall. A lifebuoy and line should still be kept ready at a close position.

15.2.6   Gantlines should be kept clear of sharp edges.

15.2.7   The anchoring points for lines, blocks and lizards must be of adequate strength and, where practicable, be permanent fixtures to the ship's structure. Integral lugs should be hammer tested. Portable rails or stanchions should not be used as anchoring points. Beam clamps and similar devices should be used solely for their intended purposes and then only under close supervision.

15.2.8   Stages and staging which are not suspended should always be secured against movement. Hanging stages should be restricted against movement to the extent practicable.

15.2.9   In machinery spaces, staging and its supports should be kept clear of contact with hot surfaces and moving parts of machinery. In the engine room, a crane gantry should not be used directly as a platform for cleaning or painting, but can be used as the base for a stable platform if the precautions of Paragraph 14.1.6 are taken.

15.2.10   Where men working from a stage are required to raise or lower themselves, great care must be taken to keep movements of the stage small and closely controlled.

15.3 Bosun's chair

15.3.1   When used with a gantline the chair should be secured to it with a double sheet bend and the end seized to the standing part with adequate tail.

15.3.2   Hooks should not be used to secure bosun's chairs unless they are of the type which because of their special construction cannot be accidentally dislodged.

15.3.3   On each occasion that a bosun's chair is rigged for use, the chair, gantlines and lizards should be thoroughly examined, and a load test applied before a man is hoisted.

15.3.4   When a chair is to be used for riding topping lifts or stays, it is essential that the bow of the shackle, and not the pin, rides on the wire. The pin in any case should be seized.

15.3.5   When it is necessary to haul a seaman aloft in a bosun's chair it should be done only by hand; a winch should not be used.

15.3.6   If a man is required to lower himself while using a bosun's chair, he should first frap both parts of the gantline together with a suitable piece of line to secure the chair before making the lowering hitch. The practice of holding on with one hand and making the lowering hitch with the other is dangerous.

15.4 Ropes

15.4.1   The safety of the man aloft or overside depends upon the strength of the line holding him, whether it is a lifeline to his harness or gantline to a bosun's chair or stage.

15.4.2   Many types of rope of both man-made and natural fibre are available, each with different properties and with different resistance to contamination by substances in use about the ship which may seriously weaken the rope. Guidance on the selection of man-made fibres can be found in the relevant British Standard (see also section 16.2). Seafarers should therefore be aware of the general limitations of the different types of rope and the following table is set out as a guide on the resistance of the main rope types to chemical attack. This table is indicative only of the

BS 4128 1967

	Resistance to chemicals of rope made of			
Substance	Manila or Sisal	Polyamide (nylon)	Polyester	Polypropylene
Sulphuric (battery) acid	None	Poor	Good	V Good
Hydrochloric acid	None	Poor	Good	V Good
Typical rust remover	Poor	Fair	Good	V Good
Caustic soda	None	Good	Fair	V Good
Liquid bleach	None	Good	V Good	V Good
Creosote, crude oil	Fair	None	Good	V Good
Phenols, Crude tar	Good	Fair	Good	Good
Diesel oil	Good	Good	Good	Good
Synthetic detergents	Poor	Good	Good	Good
Chlorinated solvents, eg trichloroethylene (used in some paint and varnish removers)	Poor	Fair	Good	Poor
Other organic solvents	Good	Good	Good	Good

possible extent of deterioration of rope; in practice, much depends upon the precise formulation of the material, the amount of contamination the rope receives and the length of time and the temperature at which is it exposed to the contamination. In some cases, damage may not be apparent even on close visual inspection.

15.4.3   Ropes should be stored away from heat and sunlight, and in a separate compartment from containers of chemicals, detergents, rust removers, paint strippers or other substances capable of damaging them.

15.4.4   The person responsible for the work being undertaken should ensure that all ropes, lifelines, gantlines etc employed for a particular job are resistant to attack by substances that might be used during the course of that job. Ropes of natural fibres, or a mixture of natural and man-made fibres, should not be used for these purposes. Similarly, care should be taken in the selection and use of ancillary equipment such as safety harnesses and safety nets.

15.4.5   Polypropylene ropes which have the best all round resistance to attack by harmful substances are generally preferred but unless they are of a type resistant to actinic degradation, such as those approved for life-saving appliances by the Department of Transport, they should not be exposed to strong sunlight for long periods. They should also be of a type providing grip comparable to that of manila or sisal ropes.

15.4.6   Rope of man-made material stretches under load to an extent which varies according to the material. Polyamide rope stretches the most.

15.4.7   Rope should be inspected internally and externally before use for signs of deterioration, undue wear or damage. This is particularly important if a gantline has not been used for some time. A high degree of powdering between strands of man-made fibre ropes indicates hard wear and impaired strength: the internal wear will be greater with ropes that stretch. Some ropes, for example of polyamide, become stiff and hard when overworked.

15.4.8   Before use, lifelines and gantlines, lizards and chairs should be load-tested to four or five times the loads they will be required to carry.

15.4.9   Some superficial splashing or wetting of lines by corrosive or rotting substances may be unavoidable during the progress of the work. The ropes etc chosen should not be susceptible to damage by the contaminant (see table above) and it should be sufficient to ensure that any effects of contamination are examined as soon as possible but in any case at the end of a day's work.

15.4.10   Mildew does not attack man-made fibre ropes but moulds can form on them. This will not affect their strength.

15.4.11   Eye or loop splices in ropes of polyamide or polyester materials should be made with four full tucks each with the completed strands of the rope followed by two tapered tucks for which the strands are halved and quartered for one tuck each respectively. Those portions of the splices containing the tucks with the reduced number of filaments should be securely wrapped with adhesive tape or other suitable material. Splices in polypropylene ropes should have at least three full tucks. The length of the splicing tails protruding from the finished splice should equal not less than three rope diameters.

15.4.12   Mechanical fastenings should not be used in lieu of splices on man-made fibre ropes because strands may be damaged during application of the mechanical fastening and the grip of the fastenings may be much affected by slight unavoidable fluctuations in the diameter of strands.

15.5   Portable ladders

15.5.1   A portable ladder should have a clear width of at least 255 mm (10 inches), be soundly constructed and have adequate strength for the purpose for which it is used. A ladder should not be used if any part is defective, for example if any rung depends for support solely on nails, spikes or similar improvisations.

15.5.2   All ladders should be inspected at regular intervals and maintained in sound condition. Wooden ladders should not be painted or treated so as to hide cracks and defects.

15.5.3   When not in use, portable ladders should be stowed in a dry ventilated space away from heat.

15.5.4   A ladder in use should rise to a height of at least 1 metre (39 inches) above the top landing place unless these are other suitable handholds.

15.5.5   A portable ladder, whether rope or rigid type, must be adequately secured against displacement as near as possible to its upper resting place.

15.5.6   There should be a clearance of at least 150 mm behind all rungs.

15.5.7   Rigid portable ladders should be pitched at a safe angle between 65° and 70° to the horizontal (ie a slope of about one horizontal for four vertical). They should stand on a firm base and be lashed in position.

15.5.8   Planks should not be supported on the rungs of portable ladders to be used as a staging, nor should ladders be used horizontally for the same purpose.

15.5.9   A man negotiating a ladder needs both hands free; he should not attempt to carry tools or equipment in his hands. If he is wearing gloves or his hands are greasy, he must take extra care.

15.5.10   Working from ladders should be avoided as far as practicable since there is a risk of overbalancing and falling. Where it is necessary, a safety harness with a lifeline secured above the position of work should be worn when working at a height in excess of 2 metres (6½ feet) (see 15.1.3).

CHAPTER 16

Anchoring, making fast, casting off and towing

16.1 Anchoring

16.1.1   Before anchors are let go, a check should be made that there are no small craft or obstacles under the bow.

16.1.2   The man operating the brake and others in the vicinity should wear safety goggles and safety helmets to avoid the risk of injuries from dirt and rust particles and debris thrown off as the chain pays out.

16.1.3   Instructions given by portable transceivers (walkie-talkies) should always be identified with the ship, preferably by including her name in the instruction.

16.1.4   Seamen engaged in stowing an anchor cable into the locker should stand in a protected position and as far as practicable should keep in constant communication with the windlass operator.

16.1.5   Anchors housed and not required should be properly secured to guard against accidents or damage should the windlass brake be released inadvertently.

16.2 Characteristics of man-made fibre ropes

16.2.1   Safe handling of man-made fibre ropes requires techniques which differ from those for handling natural fibre ropes (see also section 15.4).

16.2.2   Man-made fibre ropes are relatively stronger than those of natural fibre and so for any given breaking strain have appreciably smaller circumferences, but wear or damage will diminish strength to a greater extent than would the same amount of wear or damage on a natural fibre rope. Recommendations for substitution of natural fibre ropes by man-made fibre ropes are given in the following table:

Manila		Polyamide (Nylon) etc		Polyester (Terylene etc)		Polypropylene	
Dia	Size	Dia	Size	Dia	Size	Dia	Size
48 mm	(6)	48 mm	(6)	48 mm	(6)	48 mm	(6)
56 mm	(7)	48 mm	(6)	48 mm	(6)	52 mm	(6½)
64 mm	(8)	52 mm	(6½)	52 mm	(6½)	56 mm	(7)
72 mm	(9)	60 mm	(7½)	60 mm	(7½)	64 mm	(8)
80 mm	(10)	64 mm	(8)	64 mm	(8)	72 mm	(9)
88 mm	(11)	72 mm	(9)	72 mm	(9)	80 mm	(10)
96 mm	(12)	80 mm	(10)	80 mm	(10)	88 mm	(11)
112 mm	(14)	88 mm	(11)	88 mm	(11)	96 mm	(12)

Diameter given for 3-strand, size no for 8-strand plaited.

Polyamide (nylon) or polyester ropes are recommended for mooring tails and should have a minimum finished length of 11 metres (6 fathoms) in order to provide the necessary elasticity. The strength of the tail when new should be approximately 25 per cent greater than that of the mooring rope or wire.

16.2.3 Man-made fibre ropes have high durability and low water absorption and are very resistant to rot or mildew but should not be left unduly long exposed to sunlight when not in use. They should be covered by tarpaulins or, if the ship is on a long voyage, stowed away.

16.2.4 Ropes should be kept free of contamination by chemicals (rust removers and paint strippers may be particularly damaging) and not stowed close to any source of heat. Any accidental contamination should be reported immediately for cleansing or other action to be taken (see also section 15.4).

16.2.5 Careful inspection of man-made fibre ropes for wear externally and internally is necessary. A high degree of powdering observed between strands indicates excessive wear and reduced strength. Ropes with high stretch suffer greater interstrand wear than others. Hardness and stiffness in some ropes, polyamide (nylon) in particular, may also indicate overworking.

16.2.6 Man-made fibre stoppers of like material (but not polyamide) should be used on man-made fibre mooring lines, preferably using the 'West Country' method (double and reverse stoppering).

16.2.7 Unlike natural fibre ropes, man-made fibre ropes give little or no audible warning of approaching breaking point.

16.2.8 Stretch imparted to man-made fibre rope, which may be up to double that of natural fibre rope, is usually recovered almost instantaneously when tension is released. A break in the rope may therefore result in a dangerous back-lash and an item of running gear breaking loose may be projected with lethal force. Snatching of such ropes should be avoided; where it may occur inadvertently, personnel should stand well clear of the danger areas. The possibility of a mooring or towing rope parting under load is reduced by proper care, inspection and maintenance and by its proper use in service.

16.2.9 Man-made fibre ropes may easily be damaged by melting if frictional heat is generated during use. Too much friction on a warping drum may fuse the rope with consequential sticking and jumping of turns, which can be dangerous. Polypropylene is more liable to soften than other material. To avoid fusing, ropes should not be surged unnecessarily on winch barrels. For this reason, a minimum of turns should be used on the winch barrel; three turns are usually enough but on whelped drums one or two extra turns may be needed to ensure a good grip; these should be removed as soon as practicable.

16.2.10 The method of making eye splices in ropes of man-made fibres should be chosen according to the material of the rope:
(a) Polyamide (nylon) and polyester fibre ropes need four full tucks in the splice each with the completed strands of the rope followed by two tapered tucks for which the strands are halved and quartered for one tuck each respectively. The length of the splicing tail protruding from the finished splice should be equal to at least three rope diameters. The portions of the splice containing the tucks with reduced number of filaments should be securely wrapped with adhesive tape or other suitable material;

(b) polypropylene ropes should have at least three but not more than four full tucks in the splice. The protruding spliced tails should be equal to three rope diameters at least;

(c) polyethylene ropes should have four full tucks in the splice with protruding tails of three rope diameters at least.

16.3 Making fast

16.3.1 Surfaces of fairleads, bollards, bitts and drum ends should be kept clean and maintained in good condition. Rollers and fairleads should turn smoothly and a visual check be made that corrosion has not weakened them.

16.3.2 Mooring decks should have anti-slip surfaces provided by fixed treads or by treatment with anti-slip paint.

16.3.3 A mooring rope should be examined frequently throughout its length for both external wear and wear between strands. Splices should be intact.

16.3.4 Wire ropes should be regularly treated with suitable lubricants (see also 17.2.39).

16.3.5 New rope, 3-strand fibre rope and wire should be taken out of a coil in such fashion as to avoid disturbing the lay of the rope.

16.3.6 When wire is joined to a natural or man-made fibre rope, a thimble or other device should be inserted in the eye of the fibre rope; both wire and rope should have the same direction of lay.

16.3.7 A wire rope should not be used directly from a reel, unless designed for the purpose, because if the wire fouls the reel, both the reel and frame may be torn from the deck and cause injuries. Sufficient slack rope should be taken off a reel to cover all contingencies of use and be flaked out on deck in a safe manner. If there is doubt of the amount required, then the complete wire should be removed from the reel. Paying out of the wire rope should be controlled by turns round bitts or drum ends.

16.3.8 When cargo winches are used for handling springs on the main deck, suitable leads should be provided. If it is necessary to use a snatch block, precautions should be taken against its breaking loose.

16.3.9 A watchman, rigger or boatman should normally be employed to heave mooring ropes ashore. No seaman should go ashore other than by safe means.

16.3.10 A sufficient number of men must always be available at each end of the ship during mooring operations.

16.3.11 A seaman should not in any circumstances stand in a bight of rope nor, whenever avoidable, in the bight formed between the drum and the fairlead.

16.3.12 An experienced seaman should be at the winch controls throughout the whole time of the mooring operation.

16.3.13 When ropes and wires are under strain, as in towing, all persons should remain in positions of safety to the fullest possible extent.

16.3.14   Immediate action should be taken to reduce the load should any part of the system appear to be under excessive strain. Care is needed to see that ropes or wires will not jam when they come under strain, so that if necessary they can quickly be slackened off.

16.3.15   The safe method of heaving by means of turns on a drum is for one man to be stationed at the drum end with a second man backing and coiling down the slack as it is taken in.

16.3.16   Sharp-angled leads of rope or wire should always be avoided.

16.3.17   A wire should never be led across a fibre rope on a bollard. Wires and ropes should be kept in separate fairleads or bollards.

16.3.18   Wire on the drum end of a winch should not be used as a check wire.

16.3.19   When a wire is used as a slip wire, the eyes should be seized.

16.3.20   Chain stoppers should be used for stoppering off wire mooring ropes. They should be applied with two half-hitches in the form of a cow hitch suitably spaced with the tail backed against the lay of the wire to ensure that the chain neither jams nor opens up the lay of the wire.

16.4   Mooring to buoys

16.4.1   Where mooring to buoys is undertaken from a ship's launch or boat, seamen engaged in the operation should wear lifejackets and a lifebuoy with attached lifeline should be kept readily available in the boat.

16.4.2   Means should be provided to enable a man who has fallen into the water to climb back on board the launch or boat. If a boarding ladder with flexible sides is used, it should be weighted so that the lower rungs remain below the surface.

16.4.3   Where mooring to buoys is undertaken directly from the ship, a lifebuoy with attached line of sufficient length should be available for immediate use.

16.4.4   When slip wires are used for mooring to buoys or dolphins, the eyes of the wires should never be put over the bitts.

16.5   Casting off

16.5.1   A sufficient number of men should always be available at each end of the ship during casting off.

16.5.2   When casting off, persons should keep well clear of the bight of the mooring rope.

16.6   Towing

16.6.1   Towing operations may result in excessive loads being applied to ropes, fairleads, bitts and connections. A sudden failure of any element in the towing arrangements may cause death or serious injury to persons in the vicinity. The consequences of the failure of any element of the towing arrangement should be carefully considered and effective safety precautions taken.

16.6.2 The equipment used for towing should be adequately maintained and inspected before use to ensure that it is suitable for the proposed towing operation.

16.6.3 The rope used for a tow should be of adequate strength and free of defects and excessive wear.

16.6.4 Non-essential persons should keep well clear of the towing area.

16.6.5 Persons involved in a towing operation should be adequately briefed in their duties and the safety precautions to be taken.

16.6.6 Persons involved in a towing operation should wear suitable protective clothing. Hard hats should be worn to reduce the risk of head injury from heaving lines and other ropes.

16.6.7 Care should be taken to keep clear of the bights of rope and the whiplash area if a tow rope breaks.

16.6.8 When letting go a tow line persons should keep well clear of the eye which should be lowered under the control of a messenger to reduce the risk of injury to the persons involved in the towing operation.

16.6.9 Suitable means of communication should be provided between the Master and the deck crew to facilitate the relay of operational and safety instructions.

16.6.10 Suitable means of communication should be provided between the ship under tow and the tug.

MSN Nos. M718, M748, M1147, M1406 etc.

16.6.11 Further recommendations on towing are contained in Merchant Shipping Notices.

CHAPTER 17
Lifting Plant

17.1 Introduction

MS (Hatches and Lifting Plant) Regs SI 1988 No. 1639 Regulation 5

17.1.1 The Merchant Shipping (Hatches and Lifting Plant) Regulations 1988, regulations 5-10, deal with the use, handling and testing of lifting plant aboard ship. In carrying out the obligations contained in these regulations full account must be taken of the principles and the guidance described in this Chapter (Regulation 5).

17.1.2 In interpreting this Chapter, proper account should be taken of any relevant British Standards or other national and international standards giving an equivalent degree of safety.

17.2 Use of lifting plant

Regulation 6

17.2.1 The employer and the Master shall ensure that any lifting plant (ie lifting appliance plus lifting gear) used on board ships shall be of good design, of sound construction and material, of adequate strength for the purpose for which it is used, free from patent defect, properly installed or assembled and properly maintained (Regulation 6(1)).

17.2.2 In deciding whether a lifting appliance is of adequate strength for the purpose for which it is to be used, account should be taken of the weight of the associated lifting gear, and whether the gear is likely to impose additional stresses by virtue of the nature of the operation, eg grab work.

17.2.3 The requirement for maintenance means that the lifting plant should be kept in good working order, in an efficient state and in good repair. Systematic preventive maintenance should be undertaken, with due account taken of any manufacturers' instructions, which should include regular routine inspection by a person who is competent to assess whether the lifting plant is safe for continued use. These inspections which are separate from, and additional to, those required under regulation 8 should be at intervals related to the character and usage of the plant. Safety devices fitted to lifting appliances should be checked by the operator before work starts and at regular intervals thereafter to ensure that they are working properly.

17.2.4 The Master shall ensure that any one trip sling, preslung cargo sling, or any pallet or similar piece of equipment for supporting loads or lifting attachment which forms an integral part of the load shall not be used unless it is of good construction, of adequate strength for the purpose for which it is used and free from patent defect (Regulation 6(2)).

17.2.5 The employer and the Master shall ensure that a ship's lifting plant shall not be used other than in a safe and proper manner (Regulation 6(3)).

17.2.6 The employer and the Master shall ensure that except for the purpose of carrying out a test, ship's lifting plant is not loaded in excess of its safe working load (Regulation 6(4)).

17.2.7 Except in the event of an emergency endangering health or safety, no person shall operate a ship's ramp or a retractable car deck unless authorised to do so by a responsible ship's officer (Regulation 4(5)).

No person shall operate any ship's lifting plant unless he is trained and competent to do so and has been authorised by a responsible ship's officer (Regulation 6(5)).

17.2.8 Training should consist of theoretical instruction to the extent necessary to enable the trainee to appreciate the factors affecting the safe operation of the lifting plant, ship's ramp or retractable car-deck and of practical work with the appropriate plant etc under supervision. Equivalent training is also advised and authorisation required for operators of ship's ramps and retractable car decks that are not used as lifting plant.

17.2.9 For persons under 18 years of age undergoing training, the degree of direct supervision required should be related to the trainee's experience, perceived competence and the nature of the appliance etc on which he is being trained. Any work he carries out should be part of his training.

17.2.10 After training each person should undergo a test, and if he passes, should be given a certificate specifying the type of appliance on which the test was carried out. (Note: A certificate issued by the employer for this purpose should identify the employer and the vessel and include the full names of the trainee, the date of the test and the type or types of appliance. It should state that the trainee is considered competent to operate plant of the same type when authorised to do so. However standard seafarers' certificates of competency will suffice in respect of the authorised use of small non-powered lifting appliances such as handy billies; and in the case of non-seafarers (stevedores, maintenance workers etc) a written undertaking may at the Master's discretion be accepted from the employers specifying the relevant type or class of lifting appliance and stating that only competent personnel are employed to work on such plant).

17.2.11 Where authorised persons were regularly operating a class of lifting plant for a period of at least two years before 1 January 1989 they would have been considered competent for the award of a certificate from that date, provided there was no reason to believe otherwise.

17.2.12 Employers should keep records of training and testing undertaken and should ensure the routine monitoring of the competence of those operating lifting appliances.

17.2.13 Lifting appliances shoud be:—
(a) securely anchored, or
(b) adequately ballasted or counterbalanced, or
(c) supported by outriggers,
as necessary to ensure their stability when lifting.

17.2.14 If counterbalance weights are moveable, effective precautions should be taken to ensure that the lifting appliance is not used for lifting in an unstable condition. In particular all weights should be correctly installed and positioned.

17.2.15 Controls of lifting appliances should be permanently and legibly marked with their function and their operating directions shown by arrows or other simple means, indicating the position or direction of movement for hoisting or lowering, slewing or luffing, etc.

17.2.16   Make-shift extensions should not be fitted to controls nor any unauthorised alterations made to them. Foot-operated controls should have slip-resistant surfaces.

17.2.17   No lifting appliance should be used with any locking pawl, safety attachment or device rendered inoperative. If, exceptionally, limit switches need to be isolated in order to lower a crane to its stowage position, the utmost care should be taken to ensure the operation is completed safely.

17.2.18   A powered appliance should always have a person at the controls while it is in operation; it should never be left to run with a control secured in the ON position.

17.2.19   If any powered appliance is to be left unattended with the power on, loads should be taken off and controls put in "neutral" or "off" positions. Where practical, controls should be locked or otherwise inactivated to prevent accidental restarting. When work is completed, power should be shut off.

17.2.20   The person operating any lifting appliance should have no other duties which might interfere with his primary task. He should be in a proper and protected position, facing controls and, so far as is practicable, with a clear view of the whole operation.

17.2.21   Lifting appliances with pneumatic tyres should not be used unless the tyres are in a safe condition and inflated to the correct pressures. Means to check this should be provided.

17.2.22   Loads should if possible not be lifted over a person or any access way.

17.2.23   No person should be lifted by lifting plant except where the plant has been designed or especially adapted and equipped for lifting persons or for rescue or in similar emergencies.

17.2.24   Where the operator of a lifting appliance does not have a clear view of the whole of the path of travel of any load carried by that appliance, appropriate precautions should be taken to prevent danger. Generally this requirement should be met by the employment of a competent and properly trained signaller designated to give instructions to the operator. A signaller includes any person who gives directional instructions to an operator while he is moving a load, whether by manual signals, by radio or otherwise.

17.2.25   The signaller should have a clear view of the path of travel of the load where the operator of the lifting appliance cannot see it.

17.2.26   Where necessary, additional signallers should be employed to give instructions to the first signaller.

17.2.27   Every signaller should be in a position that is:—
(a) safe; and
(b) in plain view of the person to whom he is signalling unless an effective system of radio or other contact is in use.

17.2.28   All signallers should be instructed in and should follow a clear code of signals, agreed in advance and understood by all concerned in the operation. Examples of hand signals recommended for use on ships are shown in Appendix 4A.

17.2.29 If a load can be guided by fixed guides, or by electronic means, or in some other way, so that it is as safely moved as if it was being controlled by a competent team of driver and signallers, signallers will not be necessary.

17.2.30 All loads should be properly slung and properly attached to lifting gear, and all gear properly attached to appliances.

17.2.31 The use of lifting appliances to drag heavy loads with the fall at an angle to the vertical is inadvisable because of the friction and other factors involved and should only take place in exceptional circumstances where the angle is small, there is ample margin between the loads handled and the safe working load of the appliance, and particular care is taken. In all other cases winches should be used instead. Derricks should never be used in union purchase for such work.

17.2.32 Any lifts by two or more appliances simultaneously can create hazardous situations and should only be carried out where necessary. They should be properly conducted under the close supervision of a responsible person, after thorough planning of the operation. Section 7 of this Chapter provides guidance on the use of derricks in union purchase.

17.2.33 Lifting appliances should not be used in a manner likely to subject them to excessive over-turning moments.

17.2.34 Ropes, chains and slings should not be knotted.

17.2.35 A thimble or loop slice in any wire rope should have at least three tucks with a whole strand of the rope and two tucks with one half of the wires cut out of each strand. The strands in all cases should be tucked against the lay of the rope. Any other form of splice which can be shown as efficient as the above can also be used.

17.2.36 Lifting gear should not be passed around edges liable to cause damage without appropriate packing.

17.2.37 Where a particular type of load is normally lifted by special gear, such as plate clamps, other arrangements should not be substituted unless they are equally safe.

17.2.38 The manner of use of natural and man-made fibre ropes, magnetic and vacuum lifting devices and other gear should take proper account of the particular limitations of the gear and the nature of the load to be lifted.

17.2.39 Wire ropes should be regularly inspected and treated with suitable lubricants. These should be thoroughly applied so as to prevent internal corrosion as well as corrosion on the outside. The ropes should never be allowed to dry out.

17.2.40 Lifting operations should be stopped if wind conditions make it unsafe to continue them.

17.2.41 Cargo handling equipment that is lifted onto or off ships by crane or derrick should be provided with suitable points for the attachment of lifting gear, so designed as to be safe in use. The equipment should also be marked with its own gross weight and safe working load.

17.2.42 Before any attempt is made to free equipment that has become jammed under load, every effort should first be made to take off the load safely. Precautions should be taken to guard against sudden or unexpected freeing. Others not directly engaged in the operation should keep in safe or protected positions.

17.2.43 A load greater than the safe working load may be applied to lifting plant only for the purpose of a test required by the competent person under Regulation 7.

17.2.44 In the case of a single sheave block used in double purchase the working load applied to the wire should be assumed to equal half the load suspended from the block.

17.2.45 A mass in excess of the safe working load should not be lifted unless:—
(a) a test is required under Regulation 7; and
(b) the weight of the load is known and is the appropriate proof load; and
(c) the lift is a straight lift by a single appliance; and
(d) the lift is supervised by the competent person who would normally supervise a test and carry out a thorough examination; and
(e) the competent person specifies in writing that the lift is appropriate in weight and other respects to act as a test of the plant, and agrees to the detailed plan for the lift; and
(f) no person is exposed to danger thereby.

17.2.46 Any grab fitted to a lifting appliance should be of an appropriate size, taking into account the safe working load of the appliance, the additional stresses on the appliance likely to result from the operation, and the material being lifted.

17.2.47 The safe working load of a lift truck means its actual capacity, which relates the load which can be lifted to, in the case of a fork lift truck, the distance from the centre of gravity of the load from the heels of the forks. It may also specify lower capacities in certain situations, eg for lifts beyond a certain height.

17.3 Use of winches and cranes

17.3.1 The drum end of wire runners or falls should be secured to winch barrels or crane drums by proper clamps or U-bolts. The runner or fall should be long enough to leave at least three turns on the barrel or drum at maximum normal extension.

17.3.2 Slack turns of wire or rope on a barrel or drum should be avoided as they are likely to pull out suddenly under load.

17.4 Winches

17.4.1 When a winch is changed from single to double gear or vice versa, any load should be first released and the clutch should be secured so that it cannot become disengaged when the winch is working.

17.4.2 Steam winches should be so maintained that the operator is not exposed to the risk of scalding by leaks of hot water and steam.

17.4.3 Before a steam winch is operated, the cylinders and steam pipes should be cleared of water by opening the appropriate drain cocks. The stop valve between winch and deck steam line should be kept unobstructed. Adequate measures should be taken to prevent steam obscuring the driver's vision in any part of a working area.

17.5 Cranes

17.5.1 Ships' cranes should be properly operated and maintained in accordance with manufacturers' instructions and employers and masters should ensure that sufficient technical information is available including the following information:—
(a) length, size and safe working load of falls and, where appropriate, topping lifts (eg on "Thomson" cranes);
(b) safe working load of all fittings;
(c) boom limiting angles;
(d) manufacturers' instructions for replacing wires, topping up hydraulics and other maintenance as appropriate.

17.5.2 Power operated rail mounted cranes should be provided with an efficient braking mechanism which will arrest the motion along the rails.

17.5.3 The wheels of rail-mounted cranes should be provided with guards which reduce as far as possible the risk of the wheels running over persons' feet, and which will remove loose materials from the rails.

17.5.4 When a travelling crane is moved, any necessary holding bolts or clamps should be replaced before the crane is operated in its new position.

17.5.5 Access to a crane should be always by the proper means provided.

17.6 Use of derricks

17.6.1 Ships' derricks should be properly rigged and employers and masters should ensure that rigging plans are available containing the following information:—
(a) position and size of deck eye-plates;
(b) position of inboard and outboard booms;
(c) maximum headroom (ie permissible height of cargo hook above hatch coaming);
(d) maximum angle between runners;
(e) position, size and safe working load of blocks;
(f) length, size and safe working load of runners, topping lifts, guys and preventers;
(g) safe working load of shackles;
(h) position of derricks producing maximum forces (eg as shown in figure 17 of British Standard BS MA 48);
(i) optimum position for guy and preventers to resist maximum forces as at (h);
(j) combined load diagrams showing forces for a load of 1 tonne or the safe working load;
(k) guidance on the maintenance of the derrick rig.

17.6.2 The operational guidance in the remainder of this section applies generally to the conventional type of ship's derrick. For other types, such as the "Hallen" and "Stulken" derricks, the manufacturer's instructions should be followed.

17.6.3 Runner guides shoud be fitted to all derricks so that when the runner is slack, the bight is not a hazard to persons walking along the decks. Where rollers are fitted to runner guides, they should rotate freely.

17.6.4 Before a derrick is raised or lowered, all persons on deck in the vicinity should be warned so that no person stands in, or is in danger from, bights of wire and other ropes. All necessary wires should be flaked out.

17.6.5 When a single span derrick is being raised, lowered or adjusted, the hauling part of the topping lift or bull-wire (ie winch end whip) should be adequately secured to the drum end.

17.6.6 The winch driver should raise or lower the derrick at a speed consistent with the safe handling of the guys.

17.6.7 Before a derrick is raised, lowered or adjusted with a topping lift purchase, the hauling part of the span should be flaked out for its entire length in a safe manner. A seaman should back up to assist the man controlling the wire on the drum and by keeping the wire clear of turns and in making fast to the bitts or cleats. Where the hauling part of a topping lift purchase is led to a derrick span winch, the bull-wire should be handled in the same way.

17.6.8 To ensure the derrick in its final position, the topping lift purchase should be secured to bitts or cleats by first putting on three complete turns followed by four crossing turns and finally securing the whole with a lashing to prevent the turns jumping off due to the wire's natural springiness.

17.6.9 When a derrick is lowered on a topping lift purchase, a seaman should be detailed for lifting and holding the pawl bar, ready to release it should the need arise; the pawl should be fully engaged before the topping lift purchase or bull-wire is released. While employed on this duty the seaman should not attempt or be given any other task; in no circumstances should the pawl bar be wedged or lashed up.

17.6.10 A derrick having a topping winch, and particularly one that is self-powered, should not be topped hard against the mast, table or clamp in such a way that the initial heave required to free the pawl bar prior to lowering the derrick cannot be achieved in complete safety, that is, without putting an undue strain on the topping lift purchase and its attachments.

17.6.11 A heel block should be secured additionally by means of a chain or wire so that the block will be pulled into position under load but does not drop when the load is released.

17.6.12 The derrick should be lowered to the deck or crutch and properly secured whenever repairs or changes to the rig are to be carried out.

17.6.13 If heavy cargo is to be dragged under deck with ship's winches, the runner should be led directly from the heel block to avoid overloading the derrick boom and rigging. Where a heavy load is to be moved, a snatch block or bull wire should be used to provide a fair-lead for the runner and to keep the load clear of obstructions.

17.7 Use of derricks in union purchase

17.7.1 When using union purchase the following precautions should be strictly taken to avoid excessive tensions:—

(a) the angle between the married runners should not normally exceed 90° and an angle of 120° should never be exceeded;
(b) the cargo sling should be kept as short as possible so as to clear the bulwarks without the angle between the runners exceeding 90° (or 120° in special circumstances);
(c) derricks should be topped as high as practicable consistent with safe working;
(d) the derricks should not be rigged further apart than is absolutely necessary.

17.7.2   The following examples will show how rapidly excessive loads may be put on derricks, runners and attachments as the angle between runners increases:

At 60° included angle, the tension in each runner would be just over half the load; at 90° the tension would be nearly three-quarters of the load; at 120° the tension would be equal to the load; and at 175° the tension would be nearly 12 times the load.

17.7.3   When using union purchase, winch operators should wind in and pay out in step, otherwise dangerous tensions may develop in the rig.

17.7.4   An adequate preventer guy should always be rigged on the outboard side of each derrick when used in union purchase. The preventer guy should be looped over the head of the derrick, and as close to and parallel with the outboard guy as available fittings permit. Each guy should be secured to individual and adequate deck or other fastenings.

17.7.5   Narrow angles between derricks and outboard guys and between outboard guys and the vertical should be avoided in union purchase as these materially increase the loading on the guys. The angle between the outboard derrick and its outboard guy and preventer should not be too large as this may cause the outboard derrick to jack-knife. In general, the inboard derrick guys and preventer should be secured as nearly as possible at an angle of 90° to the derrick.

17.8   **Use of stoppers**

17.8.1   Where fitted, mechanical topping lift stoppers should be used. Where chain stoppers are used, they should ALWAYS be applied by two half-hitches in the form of a cow hitch suitably spaced with the remaining chain and rope tail backed round the wire and held taut to the wire.

17.8.2   A chain stopper should be shackled as near as possible in line with the span downhaul and always to an eyeplate, not passed round on a bight which would induce bending stresses similar to those in a knotted chain.

17.8.3   No stopper should be shackled to the same eyeplate as the lead block for the span downhaul; this is particularly hazardous when the lead block has to be turned to take the downhaul to the winch or secure it to bitts or cleats.

17.8.4   The span downhaul should always be eased to a stopper and the stopper should take the weight before turns are removed from the winch, bitts or cleats.

17.9 Overhaul of cargo gear

17.9.1   When a cargo block or shackle is replaced, care should be taken to ensure that the replacement is of the correct type, size and safe working load necessary for its intended use.

17.9.2   All shackles should have their pins effectively secured or seized with wire.

17.9.3   A special check should be made on completion of the work to ensure that all the split pins in blocks etc have been replaced and secured.

17.9.4   On completion of the gear overhaul, all working places should be cleaned of oil or grease.

17.10   Trucks and other mechanical handling appliances

17.10.1   Trucks for lifting and transporting should be used only by competent persons and only when the ship is in still water; they should never be used when vessels are in a seaway since the vehicles cannot be adequately controlled when the vessel is pitching and rolling.

17.10.2   Appliances powered by internal combustion engines should not be used in enclosed spaces unless the spaces are adequately ventilated. The engine should not be left running when the truck is idle.

17.10.3   When not in use or left unattended whilst the vessel is in port, trucks for lifting and transporting should be aligned along the length of the ship with brakes on, operating controls locked, and, where applicable, the forks tilted forward flush with the deck and clear of the passageway. If the trucks are on an incline, their wheels should be chocked. If not to be used for some time, and at all times whilst at sea, appliances should be properly secured to prevent movement.

17.10.4   No attempt should be made to handle a heavy load by the simultaneous use of two trucks. A truck should not be used to handle a load greater than its marked capacity or to move insecure or unsafe loads.

17.10.5   Persons other than the driver should not be carried on a truck unless it is constructed or adapted to do so. Riding on the forks of a truck is particularly dangerous.

17.10.6   The driver should be careful to keep all parts of his body within the limits of the width of the truck or load.

17.10.7   Tank containers should not be lifted directly with the forks of fork lift trucks, because of the risks of instability and of damaging the container with the ends of the forks. Tank containers may be lifted using fork lift trucks fitted with suitably designed side or top lifting attachments but care must be exercised due to the risk of surge in partly filled tanks.

17.11   Defects

17.11.1   Any defects found in any lifting plant, including plant provided by a shore authority, should be reported immediately to a responsible person who should take action appropriate to the circumstances.

17.12 Testing of lifting plant

Regulation 7

17.12.1 The employer and the Master shall ensure that no lifting plant on board ship shall be used:—
(a) after manufacture or installation, or
(b) after any repair or modification which is likely to alter the safe working load, or affect the lifting plant's strength or stability,
without first being tested by a competent person except in the case of a rope sling, manufactured from rope which has been tested by a competent person, and spliced in a safe manner (Regulation 7(1) and paragraph 17.2.35).

17.12.2 The employer and the Master shall ensure that a lifting appliance on board ship shall not be used unless it has been suitably tested by a competent person within the preceding five years (Regulation 7(2)).

17.12.3 The requirements for testing a lifting plant will be met if before use one of the following appropriate tests is carried out:—
(a) proof loading the plant concerned; or
(b) in appropriate cases by testing a sample to destruction; or
(c) in the case of re-testing after repairs or modifications, such a test that satisfies the competent prson who subsequently examines the plant (the re-testing of ships' lifting appliances may be effected by means of a static test eg by dynamometer where appropriate); or
(d) in the case of a lift truck, the test should be a functional test to verify that the truck is able to perform the task for which it was designed. This test should include a check to ensure that all controls function correctly and that all identification and capacity plates are fitted and contain correct information. A dynamic test should include travelling and manoeuvring, stacking, a lowering speed check and tilt leakage test with the rated load including relevant attachments where appropriate. Following the test the truck should be examined to ensure that it has no defects which would render it unsuitable for use.

17.12.4 Where proof loading is part of a test the test load applied should exceed the safe working load as specified in the relevant British Standard, or in other cases by at least the following:—

Proof Load (Tonnes)

SWL (Tonnes)	Lifting Appliances	Single Sheave Cargo and Pulley Blocks	Multi-Sheave Cargo and Pulley Blocks	Lifting Beams and Frames, etc	Other Lifting Gear
0–10	SWL $\times$ 1.25	SWL $\times$ 4	SWL $\times$ 2	SWL $\times$ 2	SWL $\times$ 2
11–20	SWL $\times$ 1.25	SWL $\times$ 4	SWL $\times$ 2	SWL $\times$ 1.04 + 9.6	SWL $\times$ 2
21–25	SWL + 5	SWL $\times$ 4	SWL $\times$ 2	SWL $\times$ 1.04 + 9.6	SWL $\times$ 2
26–50	SWL + 5	SWL $\times$ 4	SWL $\times$ 0.933 + 27	SWL $\times$ 1.04 + 9.6	SWL $\times$ 1.22 + 20
51–160	SWL $\times$ 1.1	SWL $\times$ 4	SWL $\times$ 0.933 + 27	SWL $\times$ 1.04 + 9.6	SWL $\times$ 1.22 + 20
161 +	SWL $\times$ 1.1	SWL $\times$ 4	SWL $\times$ 1.1	SWL $\times$ 1.1	SWL $\times$ 1.22 + 20

Note: Where a lifting appliance is normally used with a specific removable attachment and the weight of that attachment is not included in the marked safe working load as allowed in paragraph 17.14.5 of this Code then for the purposes of using the above table the safe working load of that appliance should be taken as being the marked safe working load plus the weight of the attachment.

17.12.5 To fall within the exception under Regulation 7(1), rope slings must be spliced according to appropriate British Standards or a method which can be shown to be equally as safe and efficient under all conditions of use. Ferrule-secured eye terminations are not splices and individual proof testing is required.

17.13 Examination of lifting plant

Regulation 8

17.13.1 The employer and the master shall ensure that any lifting plant shall not be used unless it has been thoroughly examined:—
(a) by a competent person at least once in every 12 month period; and
(b) following a test in accordance with Regulation 7 by the competent person who carried out that test (Regulation 8).

17.13.2 A "thorough examination" means a detailed examination by a competent person, supplemented by such dismantling as the competent person considers necessary, and access to or removal of hidden parts also at the discretion of the competent person in order to arrive at a reliable conclusion as to the safety of the plant examined.

17.13.3 The competent person may require "non-destructive testing" of lifting plant as part of any thorough examination.

17.13.4 The period of 12 months is the maximum period that must be met for the examination of all plant.

17.13.5 Where plant is subject to arduous or very frequent use more frequent thorough examinations may be appropriate: in such cases or in any other case where he thinks fit, the competent person carrying out the thorough examination may specify in his report a period of less than 12 months to the next thorough examination.

17.13.6 Any stipulations made by a competent person in his report required under Regulation 10 following examination under Regulation 8 are to be followed by the employer and the master.

17.13.7 A person chosen to act as a competent person in the examination of plant must be over 18 and have such practical and theoretical knowledge and actual experience of the type of machinery or plant which he has to examine as will enable him to detect defects or weaknesses which it is the purpose of the examination to discover and to assess their importance in relation to the strength, stability and functions of the machinery or plant.

17.14 Marking of lifting appliances and gear

Regulation 9

17.14.1 The employer and the Master shall ensure that each lifting appliance, lift truck and each item of lifting gear carried on the ship for which they are responsible shall be clearly and legibly marked with its safe working load and a means of identification, except that, in the case of lifting gear where such marking is not reasonably practicable, the safe working load shall otherwise be readily ascertainable (Regulation 9(1), and (3)).

Where the safe working load of a crane varies with its operating radius it is required to be fitted with an accurate indicator, clearly visible to the driver, showing the radius of the load lifting attachment at any time and the safe working load corresponding to that radius (Regulation 9(2)).

The employer and the Master shall ensure that each item of lifting gear which weighs a significant proportion of its own safe working load shall in addition to the requirement in Regulation 9(3) be clearly marked with its weight (Regulation 9(4)).

17.14.2   In the case of general purpose multi-legged sling assemblies, the marks should specify the safe working load at an included angle of up to 90° between:—
(a) opposite legs in a case of two-legged slings;
(b) adjacent legs in the case of three-legged slings;
(c) diagonally opposite legs in the case of four-legged slings;
and there may be a further mark of a safe working load up to a maximum such angle of 120°.

17.14.3   In the case of slings supplied in batches, a batch mark which is the same on each sling of that batch should be used as a means of identification where each sling does not have a separate individual mark of identification.

17.14.4   The requirement to mark the weight of lifting gear will generally apply to lifting beams, lifting frames, vacuum or magnetic lifting devices and other gear whose weight is substantial in relation to the loads they lift, and other gear which bears a similar relationship to the weight of the loads it is intended to be used with.

17.14.5   Where a lifting appliance is normally used with a specific removable attachment such as a clamp or spreader, the marking of the safe working load or rated capacity should specify whether the weight of that attachment is included.

17.15   Certificates and reports

Regulation 10

17.15.1   The Master shall ensure that a certificate or report in a form acceptable to the Secretary of State shall be supplied within 28 days following any test under Regulation 7 or examination under Regulation 8 and shall be kept in a safe place on board ship for a period of at least two years from receipt of the certificate or report of the next following test or examination (Regulation 10).

17.15.2   Certificates or reports are required to be written within 28 days, but when any competent person discovers a defect affecting the safety of plant he should take immediate steps to ensure that a suitable person in authority is made aware of these defects and informs the Master or his deputy, who should take appropriate action with respect to the use of the plant and the remedying of the defect.

17.15.3   Certificates or reports should be kept readily available on board and copies of the latest certificates or reports should be available to any dock worker or shore employer using the ship's plant.

17.15.4   Reports must be in a form approved by the Secretary of State. Approved forms based on the model forms prepared by the International Labour Office for the examination and testing of ships' lifting plant are shown at Appendix 4B. (Note: The model forms referred to here and in the following paragraph contain the minimum information required by ILO Convention 152. The forms produced for this purpose by the Classification Societies normally conform to this ILO requirement. However the style of the forms may be varied and additional information included providing the minimum requirement is met).

17.15.5   A register of lifting appliances and items of loose gear should be maintained in a form based on the model recommended by the ILO and shown at Appendix 4C (see note to 17.15.4).

CHAPTER 18

Hatches

18.1 Introduction

MS (Hatches and Lifting Plant) Regs SI 1988 No 1639 Regulation 4

18.1.1 The Merchant Shipping (Hatches and Lifting Plant) Regulations 1988, place an obligation on both the employer and the master of a ship to ensure that any hatch covering used on a ship shall be of sound construction and material, of adequate strength for the purpose for which it is used, free from patent defect and properly maintained (Regulation 4(2)).

18.1.2 The master shall ensure that a hatch covering shall not be used unless it can be removed and replaced, whether manually or with mechanical power, without endangering any person (Regulation 4(3)).

18.1.3 The master shall ensure that a hatch shall not be used unless the hatch covering has been completely removed, or if not completely removed, is properly secured (Regulation 4(4)).

18.1.3 The master shall ensure that a hatch shall not be used unless the hatch covering has been completely removed, or if not completely removed, is properly secured (Regulation 4(4)).

18.1.4 Except in the event of an emergency endangering health or safety, no person shall operate a hatch covering which is power-operated unless authorised to do so by a responsible ship's officer (Regulation 4(5)).

18.1.5 In carrying out these duties full account shall be taken of the principles and guidance in this Chapter (Regulation 4(5)).

18.2 General

18.2.1 Weather deck hatch covers and their securing arrangements should be inspected at regular intervals while the vessel is at sea.

18.2.2 All hatch covers should be properly maintained and defective or damaged hatch covers should be replaced or repaired as soon as possible. Damaged or defective covers should not be used, particularly during loading or unloading. All covers and beams should only be used if they are a good fit and overlap their end supports to an extent which is adequate but not excessive.

18.2.3 All weather deck hatch covers should be kept in a weather-tight condition when closed. They should be handled with care and at all times when hatches are open the area around the opening and in the hatchways should be appropriately illuminated.

18.2.4 Lifting appliances where used should be attached to hatch covers from a safe position and without a person being exposed to danger of falling or being trapped.

18.2.5 Loads should not be placed over, or work take place on, any section of hatch cover unless it is known that the cover is properly secured and can safely support the load.

18.2.6 Each member of the crew involved with the handling of hatch covers on the vessel should be properly instructed in their handling and operation. All stages of opening or closing hatches should be supervised by a ship's officer or other experienced person.

18.2.7 No hatch covering should be replaced contrary to information showing the correct replacement position.

18.2.8 Hatch covers should not be used for any other purpose.

18.3 Mechanical hatch covers

18.3.1 The appropriate manufacturer's instructions with respect to the safe operation, inspection, maintenance and repair of the type of mechanical hatch cover fitted should be complied with in all respects.

18.3.2 During operations, all persons should keep clear of the hatches and the cover stowage positions and the area should be kept clear of all items which might foul the covers or the handling equipment.

18.3.3 Special attention should be paid to the trim of the vessel when handling mechanical covers. The hatch locking pins or preventers of rolling hatch covers should not be removed until a check wire is fast to prevent premature rolling when the tracking is not horizontal.

18.3.4 Hatch wheels should be kept greased and free from dirt and the coaming runways and the drainage channels kept clean. The rubber sealing joints should be properly secured and be in such a condition as to provide a proper weathertight seal.

18.3.5 All locking and tightening devices should be secured in place on a closed hatch at all times when at sea. Securing cleats should be kept greased. Cleats, top-wedges and other tightening devices should be checked regularly whilst at sea.

18.3.6 Hatch covers should be properly secured immediately after closing or opening. They should be secured in the open position with chain preventers or by other suitable means. No one should climb on to any hatch cover unless it is properly secured.

18.4 Non-mechanical hatch covers and beams

18.4.1 Each non-mechanical hatchway should be provided with an appropriate number of properly fitting beams and hatch covers, pontoons or slab hatches adequately marked to show the correct replacement position, and with an adequate number of properly fitting tarpaulins, batten bars, side wedges and locking bars so that the hatch will remain secure and weathertight for all weather conditions.

18.4.2   Unless hatches are fitted with coamings to a height of at least 760 mm (30 inches) they should be securely covered or fenced to a height of 1 metre (39 inches) when not in use for the passage of cargo.

18.4.3   Manually handled hatch covers should be capable of being easily lifted by two men. Such hatch covers should be of adequate thickness and strength and provided with hand grips. Wooden hatch boards should be strengthened by steel bands at each end.

18.4.4   Hatch boards, hatch beams, pontoon hatches, hatch slabs and tarpaulins should be handled with care and properly stowed, stacked and secured so as not to endanger or impede the normal running of the vessel. One man should not attempt to handle hatch covers unaided unless the covers are designed for one-man operation. Hatchboards should be removed working from the centre towards the sides, and replaced from the sides towards the centre. When hauling tarpaulins seamen should walk forwards and NOT backwards so they can see where they are walking.

18.4.5   A derrick or crane being used to handle beams, pontoons or slab hatches should be positioned directly over them to lessen the risk of violent swinging once the weight has been taken.

18.4.6   Appropriate gear of adequate strength should be specially provided for the lifting of the beams, pontoons and slab hatches. Slings should be of adequate length, secured against accidental dislodgement while in use and fitted with control lanyards. The angle between arms of slings at the lifting point should not exceed 120°, in order to avoid undue stress. The winch or crane should be operated by a competent person under the direction of a ship's officer or other experienced person.

18.4.7   Beams and hatch covers remaining in position in a partly opened hatchway should be securely pinned, lashed, bolted or otherwise properly secured against accidental dislodgement.

18.4.8   Hatch covers and beams should not be removed or replaced until a check has been made that all persons are out of the hold or clear of the hatchway. Immediately before beams are to be removed, a check should be made that pins or other locking devices have been freed.

18.4.9   No one should walk out on a beam for any purpose.

18.4.10   Hatch covers should not be used in the construction of deck or cargo stages or have loads placed on them liable to damage them. Loads should not be placed on hatch coverings without the authority of a ship's officer.

18.5   Steel-hinged inspection/access lids

18.5.1   Inspection/access hatch lids should be constructed of steel or similar material, and hinged so they can be easily and safely opened or closed. Those on weather decks should be seated on watertight rubber gaskets and secured weathertight by adequate dogs, side cleats or equivalent tightening devices.

18.5.2   When not secured, inspection/access hatch lids should be capable of being easily and safely opened from above and, if practicable, from below.

18.5.3   Adequate hand grips should be provided in accessible positions to lift inspection/access hatches by hand without straining or endangering personnel.

18.5.4   Heavy or inaccessible hatch lids should be fitted with counter-weights so that they can easily be opened by one or two persons. Where a counter-weight cannot be fitted due to inaccessibility, the hatch lids should be supplied with a purchase or pulley with eye-plates or ringbolts fitted in appropriate positions so that the hatch can be opened and closed without straining or endangering personnel.

18.5.5   The hatch lids when open should be easily and safely secured against movement or accidental closing. Adequate steel hooks or other means should be provided.

CHAPTER 19
Work in cargo spaces

19.1 Access

19.1.1   Cargo spaces should always be well ventilated before entry is made. If it is necessary to enter a hold or other enclosed cargo space the precautions set out in Chapter 10 of this Code should be followed.

19.1.2   Whenever practicable, the permanent means of access should be used. In other cases, portable rigid ladders should be used (see section 15.5). When necessary, lifelines and safety harness should be available and used.

19.1.3   Should it be necessary to remove injured persons from a hold, the best available method should be adopted but where practicable all access openings should be opened and the following equipment used where available:
(a) a manually-operated davit, suitably secured over the access opening;
(b) a cage or stretcher fitted with controlling lines at the lower end.

19.1.4   When hatches are opened, there should be ample clearance for any loads which may have to be raised or lowered.

19.2 Lighting in cargo spaces

19.2.1   Cargo spaces in which work has to be undertaken should be adequately lit. Dazzle and strong contrasts of light and shadow should be avoided.

19.2.2   Open or naked lights should not be used. Portable lights, when used, should be adequately guarded and suitable for the intended purpose.

19.2.3   Portable lights should not be lowered or suspended by their cables. Leads for portable lights should be kept clear of loads, running gear and moving equipment.

19.2.4   Portable lights should be properly secured against accidental displacement.

19.2.5   Lights should not be switched off or removed before it has been ascertained that all personnel are clear of the compartment or hold.

19.3 Fencing

19.3.1   Before work is done in cargo spaces, all openings through which a person may fall should be adequately guarded or fenced (see section 9.5).

19.3.2   Guard rails should be tight with stanchions secured in position, and properly maintained.

19.3.3   Partly opened unguarded hatches should never be covered with tarpaulins; this would present a very dangerous situation which would not be apparent to a person walking across the hatch.

19.4   General precautions

19.4.1   Care should be taken when walking over dunnage which is loosely stowed or from which nails may be protruding.

19.4.2   When work is to be done near a tall stack of cargo, the cargo should be secured to prevent it falling. If the stack is of bagged cargo, damage to the bags may cause bleeding and subsequent collapse of the stow.

19.4.3   Where it is necessary to mount the face of a stow, a portable ladder should be used.

19.4.4   When work is being done on a tall stack of cargo or in places where there is a risk of falling, a safety net should be erected. It should not be secured to hatch covers.

CHAPTER 20

Work in machinery spaces

20.1 General

MS (Guarding of Machinery and Safety of Electrical Equipment) Regs SI 1988 No 1636. MSN No M.1355

20.1.1 Merchant Shipping Regulations require every dangerous part of a ship's machinery to be securely guarded unless it is so positioned or constructed that it is as safe as if it were securely guarded or is otherwise safeguarded. Guidance on the interpretation of these Regulations is given in the appropriate Merchant Shipping Notice.

20.1.2 All steam pipes, exhaust pipes and fittings which by their location and temperature present a hazard, should be adequately lagged or otherwise shielded. The insulation of heated surfaces should be properly maintained, particularly in the vicinity of oil systems.

20.1.3 Personnel required to work in machinery spaces which have high noise levels should wear suitable hearing protectors (see section 5.3).

20.1.4 Where high noise level in a machinery space or the wearing of hearing protectors may mask an audible alarm, a visual alarm of suitable intensity should be provided, where practicable, to attract attention and indicate that an audible alarm is sounding. This should preferably take the form of a light or lights with rotating reflectors.

20.1.5 The source of any oil leakage should be located and repaired as soon as practicable.

20.1.6 Waste oil should not be allowed to accumulate in the bilges or on tank tops. Any leakage of fuel, lubricating and hydraulic oil should be disposed of in accordance with Oil Pollution Regulations at the earliest opportunity. Tank tops and bilges should, wherever practicable, be painted a light colour and kept clean and well-illuminated in the vicinity of pressure oil pipes so that leaks may be readily located.

20.1.7 Great caution is required when filling any settling or other oil tank to prevent it overflowing, especially in an engine room where exhaust pipes or other hot surfaces are directly below. Manholes or other openings in the tanks should always be secured so that should a tank be overfilled the oil is directed to a safe place through the overflow arrangements.

20.1.8 Particular care should be taken when filling tanks which have their sounding pipes in the machinery spaces to ensure that weighted cocks are closed. In no case should a weighted cock on a fuel or lubricating oil tank sounding pipe or on a fuel, lubricating or hydraulic oil tank gauge be secured in the open position.

20.1.9 Engine room bilges should at all times be kept clear of rubbish and other substances so that mud-boxes are not blocked and the bilges may be readily and easily pumped.

20.1.10 Remote controls fitted for stopping machinery or pumps or for operating oil-settling tank quick-closing valves in the event of fire, should be tested regularly to ensure that they are functioning satisfactorily.

20.1.11 Cleaning solvents should always be used in accordance with manufacturers' instructions and in an area that is well ventilated.

20.1.12 Care should be taken to ensure that spare gear is properly stowed and items of machinery under overhaul safely secured so that they do not break loose and cause injury or damage even in the heaviest weather.

20.2 Boilers

20.2.1 A notice should be displayed at each boiler setting out operating instructions. Information provided by the manufacturers of the oil-burning equipment should be displayed in the boiler room.

20.2.2 To avoid the danger of a blowback when lighting boilers, the correct flashing up procedure should always be followed:
(a) there should be no loose oil on the furnace floor;
(b) the oil should be at the correct temperature for the grade of oil being used; if not, the temperature of the oil must be regulated before lighting is attempted;
(c) the furnace should be blown through with air to clear any oil vapour;
(d) the torch, specially provided for the purpose, should always be used for lighting a burner unless an adjacent burner in the same furnace is already lit; other means of ignition, such as introducing loose burning material into the furnace, should not be used. An explosion may result from attempts to relight a burner from the hot brickwork of the furnace;
(e) if all is in order, the operator should stand to one side, and the lighted torch inserted and fuel turned on. Care should be taken that there is not too much oil on the torch which could drip and possibly cause a fire;
(f) if the oil does not light immediately, the fuel supply should be turned off and the furnace ventilated by allowing air to blow through for two or three minutes to clear any oil vapour before a second attempt to light is made. During this interval the burner should be removed and the atomiser and tip inspected to verify that they are in good order;
(g) if there is a total flame failure while the burner is alight, the fuel supply should be turned off.

20.2.3 The avenues of escape from the boiler fronts and firing spaces should be kept clear.

20.2.4 Where required to be fitted, the gauge glass cover should always be in place when the glass is under pressure. If a gauge glass or cover needs to be replaced or repaired, the gauge should be shut off and drained before the cover is removed.

20.3 Unmanned machinery spaces

20.3.1 A seafarer should never enter or remain in an unmanned machinery space alone, unless he has received permission from, or been instructed by the engineer officer in charge at the time. Before entering the space, at regular intervals whilst in the space, and on having finally left the space, he must report by telephone, or other means provided, to the duty deck officer. The foregoing also applies to the engineer officer in charge. A seafarer may only be instructed to enter an unmanned machinery space alone by the designated engineer officer in charge and then he may only be sent to carry out a specific task which he may be expected to complete in a comparatively short time. Before he enters the space the method of reporting should be clearly explained and should follow the lines indicated above. On leaving the space he must also report in person to the designated engineer officer in charge. Consideration should be given in appropriate instances to using a 'permit to work' (see Chapter 7).

20.3.2   Notices of safety precautions to be observed by persons working in unmanned machinery spaces should be clearly displayed at all entrances to the space. Warning should be given that in unmanned machinery spaces there is a likelihood of machinery suddenly starting up.

20.3.3   Unmanned machinery spaces should be adequately illuminated at all times.

20.3.4   When machinery is under bridge control, the bridge should always be advised when a change in machinery setting is contemplated by the engine room staff, and before a reversion to engine room control of the machinery.

20.4   Refrigeration machinery

20.4.1   Adequate information should be available on each vessel, laying down the operating and maintenance safeguards of the refrigeration plant, the particular properties of the refrigerant and the precautions for its safe handling.

20.4.2   No one should enter a refrigerated compartment without first informing a responsible officer.

20.4.3   The compartment or flat in which refrigeration machinery is fitted should be adequately ventilated and illuminated. Where ventilation cannot be carried out efficiently by natural means, mechanical ventilation should be employed as necessary, arranged so that the machines are situated between inlets and outlets and that the air is exhausted from both the top and bottom of the compartment. Refrigerating machinery spaces within crew accommodation to which the Merchant Shipping (Crew Accommodation) Regulations apply are required to be ventilated by at least two ventilators to the open air, one of which must be fitted with an exhaust fan and have its inlet near the bottom of the space.

SIs 1978 No.795, 1979 No 491, 1984 No 41 and 1989 No 184

20.4.4   Where fitted, both the supply and exhaust fans to and from compartments in which refrigeration machinery is situated should be kept running at all times. Inlets and outlets should be kept unobstructed. When there is any doubt as to the adequacy of the ventilation, a portable fan or other suitable means should be used to assist in the removal of toxic gases from the immediate vicinity of the machine.

20.4.5   Should it be known or suspected that the refrigerant has leaked into any compartments, no attempt should be made to enter those compartments until a responsible officer has been advised of the situation. If it is necessary to enter the space, it should be ventilated to the fullest extent practicable and the person entering should wear approved breathing apparatus. A man should be stationed in constant attendance outside the space, also with breathing apparatus (See Chapter 10).

CHAPTER 21

Hydraulic and pneumatic equipment

21.1 General

21.1.1   Personnel using hydraulic and pneumatic equipment should be fully conversant with the proper procedures for its safe operation. Operating instructions should be followed at all times.

21.1.2   Operators should ensure that the system operating pressure shown on the pressure gauge is at the level recommended.

21.1.3   The equipment should not be operated if it is in any way faulty, with components that are not designed for use with the equipment, or when a safety device is missing, incorrectly adjusted or defective.

21.1.4   The equipment, if defective in any respect, should be effectively immobilised pending adjustment or repair. Only authorised personnel should undertake repairs to the equipment or adjustment of the pressure settings of safety devices (see section 22.12).

21.1.5   Prior to a hydraulic system being activated and when it is being closed down, the recommended checks should be made to ensure that there are no pockets of air or trapped pressure in the system and that there are no external leaks. Air pockets trapped in the system cause erratic action which can lead to injury or to damage to the installations or equipment.

21.1.6   Only the correct grade of hydraulic fluid should be used for topping up a hydraulic system.

21.1.7   Any spillage of hydraulic fluid should be cleared up immediately. Some fluids are based on mineral oils and any such fluid on the skin should be thoroughly washed off (see paragraph 1.2.8)

21.1.8   Where flexible hose assemblies are used, the application of a line of light coloured paint overlapping the junction of ferrule and hose will enable movement between the two to be readily noticed in advance of a failure.

21.1.9   When the equipment is in use, operators should never reach through a linkage of any hydraulically operated mechanism to set, adjust or operate the controls.

21.1.10   Before pressure is released from the system or any repairs undertaken any load should, where necessary, be adequately supported by other means. The operator should ensure that all pressures have been released before disconnecting any line, plug, valve or other component.

21.2 Hydraulic jacks

21.2.1   Jacks should be inspected before use to ensure that they are in a sound condition and that the oil in the reservoir reaches the minimum recommended level.

21.2.2   Before a jack is operated, care should be taken to ensure that it has an adequate lifting capability for the work for which it is to be used and that its foundation is level and of adequate strength.

21.2.3   Jacks should be applied only to the recommended or safe jacking points on equipment.

21.2.4   Equipment under which personnel are required to work should be properly supported with chocks, wedges or by other safe means—never by jacks alone.

21.2.5   Jack operating handles should be removed if possible when not required to be in position for raising or lowering the jack.

CHAPTER 22

Overhaul of machinery

22.1 General

22.1.1 Before any repair or maintenance work is commenced, care should be taken to ensure that all measures and precautions necessary for the safety of those concerned have been taken (see Chapter 7 on 'permit-to-work' systems).

22.1.2 No maintenance work or repair which might affect the supply of water to the fire main or sprinkler system should be started without the prior permission of the Master and Chief Engineer.

22.1.3 No alarm system should be isolated without the permission of the Chief Engineer.

22.1.4 Before machinery is serviced or repaired, measures should be taken to prevent turning or inadvertent starting as may occur with automatic or remote control systems.

22.1.5 Electrically-operated machinery should be isolated from the power supply.

22.1.6 Steam-operated machinery should have both steam and exhaust valves securely closed and, where possible, the valves locked or tied shut or some other means employed to indicate that the valves should not be opened. The same care is required when dealing with heated water under pressure as is required when working on steam-operated machinery or pipework.

22.1.7 In all cases, warning notices should be posted at or near the controls giving warning that the machinery concerned is not to be used.

22.1.8 When valves or filter covers have to be removed or similar operations have to be performed on pressurised systems, that part of the system should be isolated by closing the appropriate valves. Drain cocks should be opened to ensure that pressure is off the system.

22.1.9 When joints of pipes, fittings etc are being broken, the fastenings should not be completely removed until the joint has been broken and it has been established that no pressure remains within.

22.1.10 Before a section of the steam pipe system is opened to the steam supply, all drains should be opened. Steam should be admitted very slowly and the drains kept open until all the water has been expelled.

22.1.11 The officer in charge should give careful consideration to the hazards involved before allowing maintenance or repairs to, or immediately adjacent to, moving machinery. This should be permitted only in circumstances where no danger exists or where it is impracticable for the machinery to be stopped. The person who is to carry out the work should wear close-fitting clothing. Long hair should be covered (see 5.2.5). The officer in charge should consider whether it is necessary in the interests of safety for a second person to be in close attendance whilst the work is being carried out.

22.1.12 Heavy parts of dismantled machinery temporarily put aside should be firmly secured against movement in a seaway and, as far as practicable, be clear of walkways. Sharp projections on them should be covered when reasonably practicable.

22.1.13 Means of access to fire fighting equipment, emergency escape routes and watertight doors should never be obstructed.

22.1.14 Spare gear, tools and other equipment or material should never be left lying around, especially near to stabiliser or steering gear rams and switchboards.

22.1.15 A marline spike, steel rod, or other suitable device should be used to align holes in machinery being reassembled or mounted; fingers should never be used.

22.1.16 When guards or other safety devices have been removed from machinery to facilitate the overhaul, they should be replaced immediately the work is completed and before the machinery or equipment is tested.

22.1.17 An approved safety lamp should always be used for illuminating spaces where oil or oil vapour is present. Vapour should be dispersed by ventilation before work is done.

22.2 Protective clothing and equipment

22.2.1 Safety helmets should be worn by those engaged in the overhaul of engine room machinery where there is a risk of head injury and by others necessarily working in the area who might be struck by falling objects.

22.2.2 Seafarers required to work in machinery spaces which have high noise levels should wear suitable hearing protection.

22.2.3 Suitable eye protection should always be worn by those handling chemicals or welding, grinding, scaling, hammering, using a cold chisel or doing any other work of a similar nature.

22.2.4 If essential maintenance or repair work necessitates the removal of asbestos lagging, the precautions referred to in 1.6.4 should be adopted.

22.2.5 Spilled oil should always be cleared up immediately as a matter of habit, but floor plates will still become slippery making footholds insecure. This should be kept in mind during any work in machinery spaces where a lurch or fall could cause injury and particularly when heavy items of machinery are being handled. Risks are reduced by the wearing of suitable safety footwear with slip-resistant soles.

22.2.6 Protective clothing and equipment are described in Chapter 5.

22.3 Lifting

22.3.1   Where practicable, all items of lifting gear should be marked with safe working loads and not intentionally subjected to loads in excess of the rating. Lifting appliances which are not marked with their safe working load, should not be used (see section 17.14). Before any item of machinery is lifted its weight should be ascertained to ensure that working loads are not exceeded.

22.3.2   When machinery and, in particular, pistons are to be lifted by means of screw-in eye-bolts, the eye-bolts should be checked to ensure that they have collars, that the threads are in good condition and that the bolts are screwed hard down on to their collars. Screw holes for lifting bolts in piston heads should be cleaned and the threads checked to see that they are not wasted before the bolts are inserted.

22.4   Floor plates and handrails

22.4.1   Where provided, lifting handles should be used when a floor plate is removed or replaced. When lifting handles are not available, the plate should be levered up with a suitable tool and a chock inserted before lifting. On no account should fingers be used to prise up the edges.

22.4.2   Whenever floor plates or handrails are removed, warning notices should be posted, the openings should be effectively fenced or guarded and the area well-illuminated.

22.5   Working aloft or over bottom platforms

22.5.1   A stage or ladder should always be used when working beyond normal reach (see Chapter 15).

22.5.2   When work is done at a level above the bottom platform, precautions should be taken against heavy objects such as tools or parts of machinery falling on a person below. A firmly secured bucket or box should be used to hold tools and loose parts of machinery.

22.6   Boilers

22.6.1   Boilers should be opened only under the direction of an engineer officer. Care should be taken to check, after emptying, that the vacuum is broken before manhole doors are removed. Even if an air cock has been opened to break the vacuum, the practice should always be to loosen the manhole door nuts and break the joint before the removal of the dogs and knocking in the doors. The top manhole doors should be removed first. Personnel should stand clear of hot vapour when doors are opened.

22.6.2   No person should enter any boiler, boiler furnace or boiler flue until it has cooled sufficiently to make work in such places safe.

22.6.3   Before entry is permitted to a boiler which is part of a range of two or more boilers, the engineer officer in charge should ensure that either:
(a) all inlets through which steam or water might enter the boiler from any other part of the range have been disconnected, drained and left open to atmosphere; or, where that is not practicable;

(b) all valves or cocks, including blowdown valves controlling entry of steam or water, have been closed and securely locked, and notices posted to prevent them being opened again until authorisation is given.

The above precautions should be maintained whilst personnel remain in the boiler.

22.6.4 Every boiler, boiler furnace or boiler flue, should be adequately ventilated before anyone enters and while persons remain inside. An attendant should always be standing by outside while persons remain inside the boiler.

22.6.5 Men cleaning tubes, scaling boilers, and cleaning backends, should wear appropriate protective clothing and equipment including goggles and respirators.

22.6.6 Special care should be exercised before a boiler is entered which has not been in use for some time or where chemicals have been used to prevent rust forming. The atmosphere may be deficient in oxygen and tests should be carried out before any person is allowed to enter. See Chapter 10 for advice on entering enclosed spaces.

22.7 Auxiliary machinery and equipment

22.7.1 Before work is started on an electric generator or auxiliary machine, the machine should be stopped and the starting air valve or similar device should be secured so that it cannot be operated. A notice should be posted warning that the machine is not to be started nor the turning gear used. To avoid the danger of motoring and electric shock to any person working on the machine, it should be isolated electrically from the switchboard or starter before work is commenced. The circuit-breaker should be opened and a notice posted at the switchboard warning personnel that the breaker is not to be closed. Where practicable, the circuit-breaker should be locked open.

22.7.2 No attempt should be made to start a diesel engine without first barring round with the indicator cocks open. The barring gear should then be disengaged before starting the engine.

22.7.3 Oily deposits or flammable material should never be allowed to be present in way of diesel engine relief valves, crankcase explosion doors or scavenge belt safety discs.

22.7.4 Flammable coatings should never be applied to the internal surfaces of air starting reservoirs.

22.7.5 Care should be taken to prevent the jets from a diesel engine fuel injector impinging on the skin during testing. Leakage from other high pressure parts of injection equipment is similarly dangerous.

22.7.6 Oxygen should on no account be used for starting engines. To do so would probably cause a violent explosion.

22.8 Main engines

22.8.1 Where necessary, suitable staging, adequately secured, should be used to provide a working platform.

22.8.2   Before anyone is allowed to enter or work in the main engine crankcase or gear case, the turning gear should be engaged and a warning notice posted at the starting position.

22.8.3   Before the main engine turning gear is used, a check should be made to ensure that all personnel are clear of the crankcase and of any moving part of the main engine and that the duty deck officer has confirmed that the propeller is clear.

22.8.4   If a hot bearing has been detected in a closed crankcase, the crankcase should not be opened until sufficient time has been allowed for the bearing to cool down, otherwise the entry of air could create an explosive air/oil vapour mixture.

22.8.5   The opened crankcase or gear case should be well-ventilated to expel all flammable gases before any source of ignition, such as a portable lamp (unless of an approved safety type) is brought near to it.

22.8.6   Before the main engine is restarted, a responsible engineer officer should check that the shaft is clear and inform the duty deck officer who should confirm that the propeller is clear.

22.9   Electrical equipment

22.9.1   The risks of electric shock are much greater on board ship than they are normally ashore because the conditions of wetness, high humidity and high temperature (inducing sweating) reduce the contact resistance of the body. In those conditions, severe and even fatal shocks may be caused at voltages as low as 60V.

22.9.2   A notice of instructions on the treatment for electric shock should be posted in every space containing electrical equipment and switchgear. Immediate on the spot treatment of an unconscious patient is essential.

22.9.3   Before any work is done on electrical equipment, fuses should be removed or circuit breakers opened to ensure that all related circuits are dead. If possible, switches and circuit-breakers should be locked open or, alternatively, a 'not to be closed' notice attached (see 22.7.1). Where a fuse has been removed, it should be retained by the man working on the equipment until the job is finished. A check should be made that any interlocks or other safety devices are operative. Additional precautions are necessary to ensure safety when work is to be undertaken on equipment designed to operate at a nominal system voltage in excess of 1Kv (high voltage equipment). The work should be carried out by, or under the direct supervision of, a competent person with sufficient technical knowledge and a permit-to-work system should be operated (see Chapter 7).

22.9.4   Flammable materials should never be left or stored near switchboards.

22.9.5   Carbon tetrachloride should not be used for cleaning electrical equipment because of the high toxicity of its vapours. Other safer cleaning solvents such as 1:1:1 trichloroethane are available but, even with these, the area of use should be well-ventilated. Solvents should always be used in accordance with manufacturers' instructions.

22.9.6   Work on or near live equipment should be avoided if possible but when it is essential for the safety of the ship or for testing purposes, the following precautions should be taken.

22.9.7   A second man, who should be competent in the treatment of electric shock, should be continually in attendance.

22.9.8   The working position adopted should be safe and secure to avoid possible fatal contact with live parts arising from a slip or stumble or the movement of the vessel. Insulated gloves should be worn where practicable.

22.9.9   Contact with the deck, particularly if it is wet, should be avoided. Footwear if damp or with metal studs or rivets may give inadequate insulation. The use of a dry insulating mat at all times is recommended.

22.9.10   Contact with bare metal should be avoided. A hand-to-hand shock is especially dangerous. To minimise the risk of a second contact should the working hand accidentally touch a live part, one hand should be kept in a trouser pocket whenever practicable.

22.9.11   Wrist watches, metal identity bracelets and rings should be removed. They provide low resistance contacts with the skin. Metal fittings on clothing and footwear are also dangerous.

22.9.12   Meter probes should have only minimum amounts of metal exposed and insulation of both probes should be in good condition. Care should be taken that the probes do not short circuit adjacent connections. In measuring voltages greater than 250V the probe should be attached and removed with the circuit dead.

22.10   Refrigeration machinery and refrigerated compartments

22.10.1   No one should enter a refrigerated chamber without first informing a responsible officer (see section 20.4).

22.10.2   Personnel charging or repairing refrigeration plants should fully understand the precautions to be observed when handling the refrigerant.

22.10.3   Should it be known or suspected that the refrigerant has leaked into any compartment, no attempt should be made to enter that compartment without appropriate precautions being taken (see Chapter 10).

22.10.4   When refrigerant plants are being charged through a charging connection in the compressor suction line, it is sometimes the practice to heat the cylinder to evaporate the last of the liquid refrigerant. This should be done only by placing the cylinder in hot water or some similar indirect method and never by heating the cylinder directly with a blow lamp or other flame. Advice on the handling and storage of gas cylinders is given in section 12.7.

22.10.5   When repair or maintenance necessitates the application of heat to vessels containing refrigerant, which form component parts of the refrigeration system, it should be ensured that appropriate valves are opened to prevent build-up of pressure within the vessels.

22.11   Steering gear

22.11.1   Generally, work should not be done on steering gear when a ship is under way. If it is necessary to work on steering gear when the vessel is at sea, the ship should be stopped and suitable steps taken to immobilise the rudder by closing the valves on the hydraulic cylinders or by other appropriate and effective means.

22.12 Hydraulic and pneumatic equipment

22.12.1   Before repairs to or maintenance of hydraulic and pneumatic equipment is undertaken any load should be adequately supported by other means and all pressure in the system should be released. The part being worked upon should be isolated from the power source and a warning notice displayed by the isolating valve, which should be locked.

22.12.2   Precautions should be taken against the possibility of residual pressure being released when unions or joints are broken.

22.12.3   Absolute cleanliness is essential to the proper and safe operation of hydraulic and pneumatic systems; the working area and tools, as well as the system and its components, should be kept clean during servicing work. Care should also be taken to ensure that replacement units are clean and free from any contamination, especially fluid passages.

22.12.4   Only those replacement components which comply with manufacturers' recommendations should be used.

22.12.5   Since vapours from hydraulic fluid may be flammable, naked lights should be kept away from hydraulic equipment being tested or serviced.

22.12.6   Any renewed or replacement item of equipment should be properly inspected or tested before being put into operation within the system.

22.12.7   A jet of hydraulic fluid under pressure should never be allowed to impinge upon the skin. Any hydraulic fluid spilt on the skin should be thoroughly washed off.

22.12.8   All equipment should be in a safe condition before the system is brought into operation.

CHAPTER 23
Servicing radio and associated electronic equipment

23.1 General

23.1.1 Exposure to dangerous levels of microwave radiation should be avoided by strict adherence to instructions about special precautions contained in manufacturers' handbooks. Radar sets should not be operated with wave guides disconnected unless it is necessary for servicing purposes, when special precautions should be taken.

23.1.2 Work should not be undertaken within the marked safety radius of a Satellite Terminal Antenna unless its transmitter has been rendered inoperative.

23.1.3 Eyes are particularly vulnerable to microwave and ultraviolet radiation. Care should be taken to avoid looking directly into a radar aerial or waveguide while it is in operation or where arcing or sparking is likely to occur.

23.1.4 Exposure to dangerous levels of X-ray radiation may occur in the vicinity of faulty high voltage valves. Care should be exercised when fault tracing in the modulator circuits of radar equipment. An open circuited heater of such valves can lead to X-ray radiation where the anode voltage is in excess of 5000V.

23.1.5 Vapours of some solvents used for degreasing are toxic, particularly carbon tetrachloride which should never be used. Great care should be exercised when using solvents particularly in confined spaces; there should be no smoking. Manufacturers' instructions should be followed.

23.1.6 Some dry recorder papers used in echo sounders and facsimile recorders give off toxic fumes in use. The equipment should be well-ventilated to avoid inhalation of the fumes.

23.1.7 Radio transmitters and radar equipment should not be operated when men are working in the vicinity of aerials; the equipment should be isolated from mains supply and radio transmitters earthed. When equipment had been isolated, warning notices should be placed on transmitting and radar equipment and at the mains supply point, to prevent apparatus being switched on until clearance has been received from those concerned that they have finished the outside work.

23.1.8 Aerials should be rigged out of reach of persons standing at normal deck level or mounting easily accessible parts of the superstructure. If that is impracticable, safety screens should be erected.

23.1.9 Notices warning of the danger of high voltage should be displayed near radio transmitter aerials and lead-through insulators.

23.2 Electrical hazards

23.2.1   Conditions on board ship often create greater than normal risks of electric shock (see section 22.9). It should also be borne in mind that cuts and abrasions significantly reduce skin resistance.

23.2.2   Fuses should be removed from equipment before work is begun, and retained while the work is proceeding.

23.2.3   Where accumulators are used they should be disconnected at source; otherwise precautions should be taken to avoid short circuiting the accumulator terminals with consequent risk of burns.

23.2.4   Live chassis connected to one side of the mains are usually marked appropriately and should be handled with caution. Where the mains are AC and a transformer is interposed, the chassis is usually connected to the earth side of the supply, but this should be verified using an appropriate meter.

23.2.5   When some types of equipment are switched off but the mains switches are left on, some parts may remain live; power should always be cut off at the mains.

23.2.6   Modern equipment often embodies a master crystal enclosed in an oven; the supply to the oven is taken from an independent source and is not disconnected when the transmitter is switched off and the mains switch is off. Mains voltage will be present inside the transmitter, and care should be taken.

23.2.7   Before work is begun on the EHT section of a transmitter or other HT apparatus, with the mains switched off, all HT capacitors should be discharged using an insulated jumper, inserting a resistor in the circuit to slow the rate of discharge. This precaution should be taken even where the capacitors have permanent discharge resistors fitted.

23.2.8   An electrolytic capacitor that is suspect, or shows blistering, should be replaced, since it is liable to explode when electrical supply is on. There is a similar risk when an electrolytic capacitor is discharged by a short circuit.

23.2.9   Work at or near live equipment should be avoided if possible but where it is essential for the safety of the ship or for testing purposes then the additional precautions described in 22.9.6–12 should be taken.

23.3 Valves and semi-conductor devices

23.3.1   Valves being removed from equipment which has recently been operating should be grasped with a heat resistant cloth; in case of large valves, eg power amplifier, OP and modulators, which reach a high temperature in operation, cooling down time should be allowed before they are removed. Severe burns can result if they touch bare skin.

23.3.2   Cathode ray tubes and large thermionic valves should be handled with care; although they implode when broken, there is still a risk of severe cuts from sharp-edged glass fragments. Some special purpose devices contain vapour or gas at high pressure, for example Trigatron, but these are usually covered with a protective fibre network to contain the glass should they explode.

23.3.3 Beryllia (beryllium oxide) dust is very dangerous if inhaled or if it penetrates the skin through a cut or abrasion. It may be present in some electronic components. Cathode ray tubes, power transistors, diodes and thyristors containing it will be usually identified by the manufacturers' information provided, but lack of such information should not be taken as a positive indication of its absence. Those heat sink washers which contain it are highly polished and look like dark brass. These items should be carefully stored in their original packings until required.

23.3.4 Physical damage to components of this kind whether they are new or defective is likely to produce dangerous dust; abrasion should be avoided, they should not be worked by tools and encapsulations should be left intact. Excessive heat can be dangerous, but normal soldering with thermal shunt is safe. Damaged or broken parts should be separately and securely packed, following the manufacturer's instructions for return or disposal.

23.3.5 Persons handling parts containing beryllia should wear protective clothing, including gloves, to prevent beryllia coming into contact with the skin. Tweezers should be used where practicable. If the skin does become contaminated with the dust, affected parts, particularly any cuts, should be cleaned without delay.

23.4 Work on apparatus on extension runners or on the bench

23.4.1 Chassis on extension runners should be firmly fixed, either by self-locking devices or by use of chocks, before any work is done.

23.4.2 Where units are awkward or heavy for one person to handle easily, assistance should be sought (see Chapter 11). Strain, rupture or a slipped disc can result from a lone effort.

23.4.3 Any chassis on the bench should be firmly wedged or otherwise secured to prevent it overbalancing or moving. Should a live chassis overbalance, no attempt should be made to grab it.

23.4.4 Sharp edges and tag connectors on a chassis can cause cuts. Should the tag be alive and the skin is pierced, the shock experienced will be out of proportion to the voltage.

23.4.5 Temporary connections should be soundly made. Flexible extension cables should have good insulation and adequate current carrying capacity.

23.5 Work with visual display units

23.5.1 Visual display units (VDUs) should be so positioned that there is sufficient room to move, as necessary, around the equipment. Care should be taken to ensure that cables and wiring do not cause a hazard by obstructing movement.

23.5.2 Operators should be given adequate individual training in the use and capabilities of VDUs. This training should be adapted to the needs and ability of the person and the type of equipment.

23.5.3 Any person using VDUs regularly or frequently and for lengthy periods should be given an eye test by a qualified person before beginning VDU work and at regular intervals thereafter. If either the eye test or examination by an ophthalmologist shows that the person needs special glasses for this work these should be provided.

23.5.4 Work with VDUs can be mentally tiring and measures should be taken to minimise the risk of headache and soreness of the eyes. Lighting should be adequate for the task, with glare and reflection cut to a minimum, and the display on screen should be clear and easy to read. When appropriate the operator should be given short rest periods away from the equipment.

23.5.5 Symptoms such as neck and arm pains may arise as the result of prolonged bad posture. The VDU operator should avoid sitting in a slumped or cramped position. He should have adequate leg room and his chair should be comfortable and stable, with adjustable seat height and back rest.

23.5.6 Exceptionally, a VDU operator taking certain forms of medication may find his working efficiency is impaired. Seafarers operating VDUs should be aware of this possibility and should seek medical advice if necessary.

23.5.7 Further guidance on the safe use of VDUs is contained in the Health and Safety Executive publication "Visual Display Units" obtainable from HMSO.

CHAPTER 24

Storage batteries

24.1 General

24.1.1 When a battery is being charged it 'gases', giving off both hydrogen and oxygen. Because hydrogen is easily ignited in concentrations ranging from 4 per cent to 75 per cent in air, battery containers and compartments should be kept adequately ventilated to prevent an accumulation of dangerous gas.

24.1.2 Smoking and any type of open flame should be prohibited in a battery compartment. A conspicuous notice to this effect should be displayed at the entrance to the compartment.

24.1.3 Lighting fittings in battery compartments should be properly maintained at all times, with protective glasses in position and properly tightened. If cracked or broken glasses cannot be replaced immediately, the electric circuit should be isolated until replacements are obtained.

24.1.4 No unauthorised modifications or additions should be made to electrical equipment (including lighting fittings) in battery compartments.

24.1.5 Portable electric lamps and tools, and other portable power tools which might give rise to sparks should not be used in battery compartments.

24.1.6 The battery compartment should not be used as a store for any materials or gear not associated with the batteries.

24.1.7 A short circuit of even one cell may produce an arc or sparks which may cause an explosion of any hydrogen present. Additionally, the very heavy current which can flow in the short circuiting wire or tool may cause burns due to rapid overheating of the metal.

24.1.8 Insulation and/or guarding of cables in battery compartments should be maintained in good condition.

24.1.9 All battery connections should be kept clean and tight to avoid sparking and overheating. Temporary clip-on connections should never be used as they may work loose due to vibration and cause a spark or short circuit.

24.1.10 Metal tools, such as wrenches and spanners, should never be placed on top of batteries as they may cause sparks or short circuits. The use of insulated tools is recommended.

24.1.11 Jewellery, watches and rings etc should be removed when working on batteries. A short circuit through any of these items will heat it rapidly and may cause a severe skin burn. If rings cannot be removed, they should be heavily taped in insulating material.

24.1.12 The battery charges and all circuits fed by the battery should be switched off when leads are being connected or disconnected. If a battery is in sections, it may be possible to reduce the voltage between cells in the work area, and hence the severity of an accidental short circuit or electric shock, by removing the jumper leads between sections before work is begun. It should be appreciated that whilst individual cell voltages may not present a shock risk, dangerous voltages can exist where numbers of cells are connected together in series. A lethal shock needs a current of only a few tens of milliamps and particular care should be exercised where the voltage exceeds 50V.

24.1.13 Battery cell vent plugs should be screwed tight while connections are being made or broken.

24.1.14 The ventilation tubes of battery boxes should be examined regularly to ensure that they are free from obstruction.

24.1.15 Lids of battery boxes should be fastened while open for servicing and properly secured again when the work is finished.

24.1.16 Batteries should be kept battened in position to prevent shifting in rough weather.

24.1.17 Alkaline and lead-acid batteries should be kept in separate compartments or separated by screens. Where both lead-acid and alkaline batteries are in use, great care should be exercised to keep apart the materials and tools used in servicing each type, as contamination of the electrolyte may cause deterioration of battery performance and mixing of the two electrolytes produces a vigorous chemical reaction which could be very dangerous.

24.1.18 Both acid and alkaline electrolytes are highly corrosive. Immediate remedial action should be taken to wash off any accidental splashes on the person or on equipment. Hands should always be washed as soon as the work has finished.

24.1.19 Batteries should always be transported in the upright position to avoid spillage of electrolyte. A sufficient number of men should be employed since the batteries are heavy and painful strains or injury can otherwise easily result (see Chapter 11).

24.2 Lead-acid batteries

24.2.1 When the electrolyte is being prepared, the concentrated sulphuric acid should be added SLOWLY to the water. IF WATER IS ADDED TO THE ACID, THE HEAT GENERATED MAY CAUSE AN EXPLOSION OF STEAM, SPATTERING ACID OVER THE PERSON HANDLING IT.

24.2.2 Goggles, rubber gloves and protective apron should be worn when acid is handled.

24.2.3 To neutralise acid on skin or clothes, copious quantities of clean fresh water should be used.

24.2.4   An eyewash bottle should be to hand in the compartment for immediate use on the eyes in case of accident. This bottle should be clearly distinguishable by touch from acid or other containers, so that it may be easily located by a person who is temporarily blinded.

24.2.5   The corrosion products which form round the terminals of batteries are injurious to skin and eyes. They should be removed by brushing, away from the body. Terminals should be protected with petroleum jelly.

24.2.6   An excessive charging rate causes acid mist to be carried out of the vents onto adjacent surfaces. This should be cleaned off with diluted ammonia water or soda solution, and affected areas then dried.

24.3   Alkaline batteries

24.3.1   The general safety precautions with this type of battery are the same as for the lead-acid batteries with the following exceptions.

24.3.2   The electrolyte in these batteries is alkaline but is similarly corrosive. It should not be allowed to come into contact with the skin or clothing. In the case of contact with the skin, the affected parts should be washed with copious quantities of clean fresh water, but if burns ensue, boracic powder or a saturated solution of boracic powder should be applied. Eyes should be washed out thoroughly with plenty of clean fresh water followed immediately with a solution of boracic powder (at the rate of one teaspoonful to ½ litre or 1 pint of water). This solution should be always readily accessible when the electrolyte is handled.

24.3.3   Unlike lead-acid batteries, metal cases of alkaline batteries remain live at all times and care should be taken not to touch them or to allow metal tools to come into contact with them.

CHAPTER 25

Work in galley, pantry and other food handling areas

25.1 Health and hygiene

25.1.1 Catering staff have a responsibility for ensuring that high standards of personal hygiene and cleanliness of the galley, pantry and mess rooms are always maintained.

25.1.2 Hands and fingernails should be washed and cleaned before food is handled. This is also most important after handling uncooked meat or fish (because this contains bacteria likely to contaminate other food), and after visiting the toilet.

25.1.3 All cuts, however small, should be reported immediately and receive first aid attention to prevent infection.

25.1.4 An open cut, burn or abrasion should be covered with a waterproof dressing.

25.1.5 Illness, rashes or spots should be reported immediately the symptoms appear.

25.1.6 A person suffering from dysentery or diarrhoea should not work in the galley, pantry or other food handing areas.

25.1.7 Catering staff should wear clean clothing when handling food and preparing meals. Handbasins, hot water, soap (preferably as soap solution from a dispenser) and hand drying facilities should be available.

25.1.8 Cleanliness of all food, crockery, cutlery, linen, utensils, equipment and storage is vital. Foodstuffs and drinking water should not be stored where germs can thrive. Similarly, deep frozen food which has been defrosted is not to be refrozen.

25.1.9 Cracked or chipped crockery and glassware should be destroyed. Foodstuffs which may have come into contact with broken glass or broken crockery should be thrown away.

25.1.10 There should be no smoking in galleys, pantries, store rooms or other places where food is prepared.

25.1.11 Crockery and glassware should not be left submerged in washing up water where it may easily be broken and cause injury. Such items should be washed up individually as should knives and any utensils or implements with sharp edges.

25.1.12 Some domestic cleaning substances contain bleach or caustic soda (sodium hypochlorite) whilst some disinfectants contain carbolic acid (phenol). These substances can burn the skin and they are poisonous if swallowed. They should be treated with caution and should not be mixed together or used at more than the recommended strength. See also the guidance in 1.2.11–13.

25.1.13 Food waste, empty food containers and other garbage are major sources of pollution and disease and should be placed in proper storage facilities safely away from foodstuffs. Their discharge into the sea is prohibited except in the circumstances specified in the relevant Merchant Shipping Notice.

MSN No M1389

25.2 Slips, falls and tripping hazards

25.2.1 A large proportion of injuries to catering staff arise from slips and falls caused by wearing unsuitable footwear; 'flip-flops', sandals, plimsolls etc are especially dangerous on greasy decks and they afford no protection to the feet from burns or scalds if hot or boiling liquids are spilt. Suitable footwear, preferably with slip-resistant soles, should be worn at all times.

25.2.2 Decks and gratings should be kept free of grease, rubbish and ice etc to obviate slipping which may result in serious injuries especially when hot liquids or glass and crockery are being carried. Any spillage should be cleared up immediately.

25.2.3 Broken glass or crockery should be cleared away with a brush and pan—never with bare hands.

25.2.4 The area of deck immediately outside the entrance to refrigerated rooms should have an anti-slip surface.

25.2.5 Care should be taken when using stairs and companionways; one hand should always be kept free to grasp the handrail.

25.2.6 Trays, crates, cartons etc should not be carried in such fashion that sills, storm steps or other obstructions in the path are obscured from view.

25.3 Galley stoves, steamboilers and deep fat fryers

25.3.1 Care should be taken in lighting oil-fired galley stoves. The following procedures should always be adopted:
(a) the inside of the furnace should be checked to see that there is no oil in it and air should be blown through to clear any oil vapour; a blowback may occur if an attempt is made to light the burner with oil or oil vapour in the furnace;
(b) the oil should be at the correct temperature for the grade of oil being used; if not, the temperature of the oil should be regulated before lighting is attempted;
(c) the torch specially provided for the purpose should always be used for lighting a burner; other means of ignition, eg by the introduction of loose burning material into the stove, should not be used. An attempt to relight a burner from the hot brickwork of the stove may result in an explosion;
(d) if all is in order, the operator should stand to one side, the lighted torch should be inserted and the fuel turned on. Care should be taken that there is not too much oil on the torch, which could drip and possibly cause a fire;

(e) if the oil does not light immediately, the fuel supply should be turned off and the furnace ventilated by allowing air to blow through for two or three minutes to clear any oil vapour before a second attempt to light. During this interval, the burner should be removed and the atomiser and tip inspected to verify that they are in good order;

(f) if while the burner is alight there is a total flame failure in the furnace the fuel supply should be completely closed off.

25.3.2 Catering staff should not attempt to repair electric or oil-fired ranges or electric microwave ovens. Defects should always be reported so that proper repairs may be made. The equipment should be kept out of use and a warning notice displayed until it has been repaired.

25.3.3 The indiscriminate use of water in hosing down and washing equipment in the galley can be very dangerous, particularly where there are electrical installations. Whenever the galley deck is washed down, power to an electric range and all electric equipment should be switched off and isolated from the supply and water kept from contact with the electrical equipment.

25.3.4 Range guard rails should be used in rough weather. Pots and pans should never be filled to the extent that the contents slip over when the ship rolls.

25.3.5 Dry cloths or pot holders should always be used to handle hot pans and dishes. Wet cloths conduct heat quickly and may scald the hands.

25.3.6 No one should be directly in front of an oven when the door is opened—the initial heat blast can cause burns.

25.3.7 The steam supply to pressure cookers, steamers and boilers should be turned off and pressure released before their lids are opened.

25.3.8 Fat should not be rendered down in ovens. If forgotten, it may overheat and catch fire. A thermostatically-controlled fryer may be used for the purpose but fat is best rendered down with a little water in a deep heavy-bottomed pot.

25.3.9 Water should never be poured into hot fat; the water turns into steam, throwing the fat considerable distances. This may cause severe burns to personnel, and possibly start a fire.

25.3.10 If fat catches fire in a container, the flames should be smothered if practicable and the container removed from the source of heat. Otherwise a suitable fire extinguisher should be used. In no circumstances should water be used.

MS (Provisions and Water) Regs SI 1989 No 102; MSN No M1373 (Appendix B, para 5(b)).

25.3.11 Merchant Shipping regulations require oils or fats used for cooking purposes to be edible, sweet, long keeping and highly refined. The flash point of the cooking medium should be no lower than 315°C (600°F).

25.3.12 Deep fat fryers should be provided with suitable safety lids which should be kept in position when the fryers are not in use.

MSN No M1022

25.3.13 To minimise the risk of fire from failure of the control thermostat all deep fat fryers should be fitted with a second thermostat set to provide a thermal cut-out as specified in the relevant Merchant Shipping Notice.

25.3.14 Electrically operated deep fat fryers should be switched off immediately after use.

25.3.15 When microwave ovens are used, and particularly with pre-frozen foods, it is important to ensure that the food is cooked thoroughly and evenly. The instructions issued by the oven manufacturers should be followed carefully in conjunction with the information on the packaging of the foodstuff.

25.3.16 No microwave oven should be operated if the oven door or its interlock is out of use, the door broken or ill fitting or the door seals damaged. Each microwave oven should carry a permanent notice to this effect.

25.4 Catering equipment

25.4.1 Except under the supervision of an experienced person, no one should use catering equipment unless trained in its use and fully instructed in the precautions to be observed.

25.4.2 Dangerous parts of catering machines should be properly guarded and the guards kept in position whenever the machine is in use.

25.4.3 Any machine or equipment that is defective in its parts, guards or safety devices should be reported and taken out of service, with power disconnected, until repaired.

25.4.4 When a power-operated machine has to be cleaned or a blockage in it removed, it should be switched off and isolated from the power supply. Some machines will continue to run down for a while thereafter, and care should be taken to see that dangerous parts have come to rest before cleaning is begun.

25.4.5 A safe procedure for cleaning all machines should be established and carefully followed. Every precaution should be taken when cutting edges, for example on slicing machines, are exposed by the necessary removal of guards to allow thorough cleaning. Guards should be properly and securely replaced immediately the job is done.

25.4.6 Unless properly supervised, a person under eighteen years of age should not clean any power operated or manually driven machine with dangerous parts which may move during the cleaning operation.

25.4.7 Appropriate implements, not fingers, should be used to feed materials into processing machines.

25.4.8 Electrical equipment should not be used with wet hands.

25.5 Knives, saws, choppers etc

25.5.1 Sharp implements should be treated with respect and handled with care at all times. They should not be left lying around working areas where someone may accidentally cut themselves. They should not be mixed in with other items for washing up but cleaned individually and should be stored in a safe place.

25.5.2 Knives should be kept tidily in secure racks or sheaths when not in use.

25.5.3 The handles of knives, saws, choppers etc should be securely fixed and kept clean and free from grease. The cutting edges should be kept clean and sharp.

25.5.4 Proper can openers in clean condition should be used to open cans; improvisations are dangerous and may leave jagged edges on the can.

25.5.5 Chopping meat requires undivided attention. The chopping block must be firm, the cutting area of the meat well on the block and hands and body clear of the line of strike. There must be adequate room for movement and no obstructions in the way of the cutting stroke. Particular care is required when the vessel is moving in a seaway.

25.5.6 Foodstuffs being chopped with a knife should not be fed towards the blade with outstretched fingers. Finger tips should be bent inwards towards the palm of the hand with the thumb overlapped by the forefinger. The knife blade should be angled away from the work and so away from the fingers.

25.5.7 A falling knife should be left to fall, not grabbed.

25.5.8 A meat saw should be guided by the forefinger of the free hand over the top of the blade. The use of firm even strokes will allow the blade to feel its way; if forced, the saw may jump possibly causing injury.

25.6 Refrigerated rooms and store rooms

25.6.1 All refrigerated room doors should be fitted with means both of opening the door and of sounding an alarm from the inside.

25.6.2 A routine testing of the alarm bell and checking of the door clasps and inside release should be carried out regularly, at least at weekly intervals.

25.6.3 Those using the refrigerated room should make themselves familiar with the operation, in darkness, of the inside release for the door and the location of the alarm button.

25.6.4 All refrigerated room doors should be fitted with an arrangement of adequate strength to hold the door open in a seaway and should be secured open while stores are being handled. These doors are extremely heavy and can cause serious injury to a person caught between the door and the jamb.

25.6.5 Anyone going into a refrigerated room should take the padlock, if any, inside with him. Another person should be informed.

25.6.6 Cold stores or refrigerated rooms should not be entered if it is suspected that there has been a leakage of refrigerant. A warning notice to this effect should be posted outside the doors.

25.6.7 All stores and crates should be stowed securely so they do not shift or move in a seaway.

25.6.8 When wooden boxes or crates are opened, protruding fastenings should be removed or made safe.

25.6.9 Where a metal strip secures the lid of a tea chest it should be completely removed or made safe, otherwise its jagged edge may cause injury.

25.6.10 Metal hooks not in use should be stowed in a special container provided for the purpose. Where hooks cannot be removed they should be kept clear.

CHAPTER 26

Work in ships' laundries

26.1 General

26.1.1 Many of the general hazards found in the ship's laundry are similar to those elsewhere on the ship; strains can be caused by improper handling of awkward or heavy loads, equipment and washing left in gangways can cause falls and defective or improperly used machinery can cause injuries. There are also specific hazards in laundries which require particular attention.

26.1.2 Floors in laundry spaces can become extremely slippery when wet, especially where soap solutions are used. Constant care is needed and suitable footwear should be worn (see 5.6.2-3 and 9.10.1(b)).

26.2 Burns and scalds

26.2.1 Many accidents in laundries involve scalding by steam or hot liquids or burns from hot surfaces. Every hot vessel or machine and every container of scalding liquid should be regarded as a potential danger, capable of causing injury and adequate precautions should be taken.

26.2.2 Good ventilation should be maintained to reduce heat and humidity in the working environment.

26.2.3 Where high pressure steam is used to heat water, the supply pressure should be regulated to ensure excessive steam cannot be applied to any machine.

26.3 Machinery and equipment

26.3.1 All personnel required to work in the laundry or use any part of the equipment there should be fully instructed on the proper operation of the machinery. A person under 18 years of age should not work on industrial washing machines, hydro-extractors, calenders or garment presses unless he is fully instructed as to precautions to be observed, and has received sufficient training in work at the machine or is under close supervision by a suitably experienced person.

26.3.2 Equipment should be inspected before use for faults and damage. Particular attention should be paid to the automatic cut-off or interlocking arrangements on washing machines, hydro-extractors etc and the guards and emergency stops on presses, calenders, mangling and wringing machines. Any defect or irregularity found during inspection, or apparent during operation of the equipment, should be reported immediately and the use of the machine discontinued until such time as any necessary repairs or adjustments have been carried out. A notice warning against use should be displayed prominently on the defective machine.

26.3.3 Frequent and regular inspection and thorough checking of all electrical equipment and apparatus are also necessary to ensure the standard of maintenance essential under the conditions which prevail in laundries.

26.3.4 Machines should not be overloaded and loads should be distributed uniformly.

26.3.5 Reliance should not be placed entirely on interlocking or cut-off arrangements on the doors of washing machines, hydro-extractors and drying tumblers etc; doors should not be opened until all movement has ceased.

26.4 Dry-cleaning operations

26.4.1 The principal hazard presented by a dry-cleaning solvent is that it is highly volatile, producing a vapour which is anaesthetic. The vapour is capable of inducing drowsiness, followed by unconsciousness and eventually death if the concentration is high enough and the affected person is not quickly removed to fresh air. Effective mechanical ventilation should therefore be provided in any compartment containing dry-cleaning plant. The purpose of such ventilation is to ensure that the vapour concentration never exceeds the 'threshold limit value' which is the airborne concentration of vapour to which it is believed that nearly all persons may be repeatedly exposed without adverse effects.

26.4.2 Another hazard is that the vapour, if allowed to contact naked flames or red-hot surfaces, decomposes into toxic and corrosive substances which are dangers to health. Smoking should therefore be prohibited in compartments where the solvent is present.

26.4.3 The vapour is heavier than air, and may therefore build up in the bottom of a compartment.

26.4.4 Dry cleaning solvent is also a potential cause of de-fatting of the skin, which may lead to cracking of the skin with the resultant possibility of infection from other sources. Suitable impermeable, cotton-lined PVC gloves should therefore be worn when handling it.

26.4.5 An officer should be appointed to take overall responsibility for the security and operation of the dry-cleaning plant. The responsible officer should ensure that the plant compartment is locked at all times when the plant is not in use, and that only the operator and himself normally has access to it.

26.4.6 The responsible officer should be satisfied that the plant operator is fully aware of:
(a) the operating and maintenance procedures as laid down in the manufacturer's instruction manual, a copy of which should be kept on board the ship;
(b) action necessary in the event of malfunctions of the plant, in the interests of personal safety;
(c) the precautions to be taken in handling the solvent, and dealing with spillages;
(d) the inherent dangers of excessive concentrations of solvent vapour;
(e) the meaning of cautionary notices displayed in the plant compartment.
Appropriate protective clothing should be made available, and a second person should be present in case of emergency.

26.4.7 Warning notices appropriate to the particular solvent used should be displayed prominently and permanently in the plant compartment. Such notices should be obtained from the solvent manufacturer, and should include instructions in first aid for persons overcome by solvent vapour.

26.4.8 Where separate compartments are provided for the storage of solvent, the same general precautions as above should be observed, including the provision of suitable warning notices and the instructions in first aid.

26.4.9 Solvent being transferred manually from the store to the plant compartment should be carried in a closed container.

26.4.10 Thick and padded articles, such as sleeping bags and quilts, should not be dry-cleaned. They are very retentive of the dry-cleaning solvent from which it is extremely difficult to free them, and body warmth when the articles are put into use again is likely to cause vapours to be emitted from the trapped residues.

26.4.11 Articles which have been dry-cleaned should be aired very throughly on the ventilated rack provided, to remove any solvent fumes.

26.5 Fire prevention

26.5.1 Hand pressing irons should not be left standing on materials likely to be combustible.

26.5.2 Clothing should be left to dry only in the designated places. It should not be put to dry in any machinery space or on or close to electric heaters, radiators, etc in accommodation spaces.

CHAPTER 27

General cargo ships

Note: Chapters 11, 17, 18 and 19 have special relevance to work on general cargo ships.

27.1 Stowage of cargo

27.1.1   Cargo should be stowed in 'tween decks with due regard to the order of discharging. When beams and hatch covers have to be removed with cargo remaining in the 'tween deck, an access space at least one metre wide should be left adjacent to any open area of hatchway. Guidelines should be painted around 'tween deck hatchways at a distance of one metre from the coamings.

27.1.2   Wherever practicable, cargo should be stowed so as to leave safe clearance behind the rungs of hold ladders and to allow safe access as may be necessary at sea.

MS (Load Line) (Deck Cargo) Regs SI 1968 No 1089

27.1.3   Deck cargo should be stowed in accordance with the relevant statutory provisions, clear of hatch coamings to leave safe access for seamen. Obstructions in the access way such as lashings or securing points should be painted white to make them more easily visible. Where this is impracticable and cargo is stowed against ships' rails or hatch coamings to such a height that the rails or coamings do not give effective protection to the crew from falling overboard or into the open hold, temporary fencing should be provided (see section 9.5).

27.1.4   When deck cargo is stowed against and above ship's rails or bulwarks, a wire rope pendant or a chain, extending from the ring bolts or other anchorage on the decks to the full height of the deck cargo, should be provided to save seamen having to go overside to attach derrick guys and preventers directly to the anchorages on the deck.

27.1.5   Where hatches will have to be opened at intermediate ports before deck cargo is unloaded, it should be stowed leaving a clear space of at least one metre around the coamings or around the part of the hatch that is to be opened. If this is impracticable, provision should be made by fencing or life-lines, to enable seamen to remove and replace beams and hatch coverings in safety.

27.2 Dangerous goods and substances

MS (Dangerous Goods and Marine Pollutants) Regs SI 1990 No 2605.

27.2.1   Merchant Shipping regulations lay down requirements for carriage of dangerous substances and the provisions of the International Maritime Dangerous Goods (IMDG) Code together with those contained in various Merchant Shipping Notices should be observed. The IMDG Code contains details of classification, documentation, packaging etc and advice on such application as will meet the requirements of the regulations. In particular it lists and gives details of many dangerous substances.

27.2.2 The general introduction and the introductions to individual classes of the IMDG Code contain many provisions to ensure the safe handling and carriage of dangerous goods including requirements for electrical equipment and wiring, fire fighting equipment, ventilation, smoking, repair work, provision and availability of special equipment etc, some of which are general for all classes and others particular to certain classes only. It is important that reference should be made to this information before handling dangerous goods. Some of the requirements are highlighted in subsequent paragraphs. Where any doubts exist, advice should be sought from the Department of Transport or other competent authority.

27.2.3 Dangerous substances should be loaded or unloaded only under the supervision of a responsible officer.

27.2.4 Dangerous substances should not be loaded other than in accordance with the regulations—ie in accordance with the IMDG Code. In the case of certain solid dangerous substances shipped in bulk, loading should be carried out in accordance with Appendix B of the Code of Safe Practice for Solid Bulk Cargoes published by the International Maritime Organization (IMO).

27.2.5 When seamen are required to handle consignments containing dangerous substances, adequate information should be available as to the nature of such substances and any special precautions to be observed. In the event of accidental exposure to dangerous substances, reference should be made to the Medical First Aid Guide for Use in Accidents Involving Dangerous Goods (MFAG) published by IMO.

27.2.6 Suitable precautions, such as the provision of special lifting gear, should be taken to prevent damage to receptacles containing dangerous substances.

27.2.7 In compartments containing cargo having an explosion or fire risk (for example, explosives and flammable liquids), all electrical circuits and equipment should meet the recommendations of the IMDG Code. When loading or unloading such cargo, firefighting equipment should be rigged ready for immediate use. Smoking should be prohibited while cargo handling is in progress, except in authorised places.

27.2.8 Where necessary, seamen loading, unloading or otherwise handling dangerous substances should wear appropriate protective clothing and personal protective equipment, including respiratory equipment (see Chapter 5).

27.2.9 Appropriate measures should be taken promptly to render harmless any spillage of dangerous substances. Particular care should be taken when dangerous substances are carried in refrigerated spaces where any spillage may be absorbed by the insulating material. Insulation affected in this way should be inspected and renewed if necessary.

27.2.10 Where there is leakage or escape of dangerous gases or vapours from the cargo, seamen should leave the danger area. The area should be ventilated and tested, if possible, to verify that the concentration of gases or vapours in the atmosphere is not high enough to be dangerous, before personnel are allowed to enter the area again. Seamen required to deal with spillages or to remove defective packages should be provided with and wear suitable breathing apparatus and protective clothing as the circumstances dictate. Suitable rescue and resuscitation equipment should be readily available in case of an emergency (see Chapter 10).

27.2.11   When dangerous goods are to be carried and in particular when incidents such as those described in 27.2.9 and 27.2.10 occur or are suspected, the Code 'Emergency Procedures for Ships Carrying Dangerous Goods', published by the IMO, should be consulted.

27.2.12   Guidance on the assessment and control of risks from substances hazardous to health is also given in paragraphs 1.5.1 to 1.5.10.

27.3 Working cargo

27.3.1   No other work such as chipping, caulking, spray painting, shotblasting or welding etc, should be carried out in a space where cargo working is in progress if it thereby gives rise to a hazard to persons working in the space.

27.3.2   Loads being lowered or hoisted should not pass or remain over any person engaged in loading or unloading or performing other work in the vicinity.

27.3.3   Care should be taken when using ladders in the square of the hatch while cargo is being worked.

27.3.4   A signaller should always be employed at a hatchway when cargo is being loaded or discharged unless the crane driver or winchman has a complete unrestricted view of the load. Guidance for signallers is given in paragraphs 17.2.24 to 17.2.29

27.3.5   The signaller should be in a position where he can best follow the work and be seen by the crane driver or winch operator. If his view is not clear and unobstructed, additional signallers should be employed to give instructions to the signaller guiding the person operating the crane or winch.

27.3.6   Before giving a signal to hoist, the signaller should receive clearance from the person making up the load that it is secure, and should ascertain that no one else would be endangered by the hoist. Before giving the signal to lower, he should warn persons in the way and ensure all are clear.

27.3.7   When a load does not ride properly after being hoisted, the signaller should immediately give warning of danger and the load should be lowered and adjusted as necessary.

27.3.8   Hooks, slings and other gear should not be loaded beyond their safe working loads.

27.3.9   Loads should be raised and lowered smoothly, avoiding sudden jerks or 'snatching' loads.

27.3.10   Strops and slings should be of sufficient size and length to enable them to be used safely and be so applied and pulled sufficiently tight to prevent the load or any part of the load from slipping and falling.

27.3.11   Loads (sets) should be properly put together and properly slung before they are hoisted or lowered.

27.3.12 Before heavy loads such as long lengths of steel sections, tubes, lumber, etc are swung, the load should be given a trial lift in order to test the efficacy of the slinging.

27.3.13 Except for the purpose of breaking out or making up slings, lifting hooks should not be attached to:
(a) the bands, strops or other fastenings of packages of cargo, unless these fastenings have been specifically provided for lifting purposes;
(b) the rims (chines) of barrels or drums for lifting purposes, unless the construction and condition of the barrels or drums is such as to permit lifting to be done safely with properly designed and constructed can hooks.

27.3.14 Suitable precautions, such as the use of packing or chafing pieces, should be taken to prevent chains, wire and fibre ropes from being damaged by the sharp edges of loads.

27.3.15 When slings are used with barrel hooks or similar holding devices where the weight of the load holds the hooks in place, the sling should be led down through the egg or eye link and through the eye of each hook in turn so that the horizontal part of the sling draws the hooks together.

27.3.16 The angle between the legs of slings should not normally exceed 90°. Where this is not reasonably practicable, the angle may be exceeded up to 120° provided that the slings have been designed to work at the greater angles.

27.3.17 Trays and pallets should be hoisted with four-legged slings and where necessary nets or other means should be used to prevent any part of the load falling.

27.3.18 When bundles of long metal goods such as tubes, pipes and rails are being hoisted, two slings should be used and, where necessary, a spreader. A suitable lanyard should also be attached, where necessary.

27.3.19 Wire rope slings of adequate size should be used for loading and discharging logs; tongs should not be used except to break out loads.

27.3.20 Cargo buckets, tubs and similar appliances should be carefully filled so that there is no risk of the contents falling out and be securely attached to the hoist (for example, by a shackle) to prevent tipping and displacement during hoisting and lowering.

27.3.21 Shackles should be used for slinging thick sheet metal, if there are suitable holes in the material; otherwise suitable clamps on an endless sling should be used.

27.3.22 Bricks and other loose goods of similar shape, carboys, small drums, canisters, etc should be loaded or discharged in suitable boxes or pallets with sufficiently high sides, lifted by four-legged slings.

27.3.23 Slings or chains being returned to the loading position should be securely hooked on the cargo hook before the signaller gives the signal to hoist. Hooks or claws should be attached to the egg link or shackle of the cargo hook, not allowed to hang loose. The cargo hook should be kept high enough to keep slings or chains clear of persons and obstructions.

27.3.24 When work is interrupted or has ceased for the time being, the hatch should be left in a safe condition, with either guard rails or the hatch covers in position.

CHAPTER 28

Carriage of containers

28.1.1   To the extent that containers are simply packages of pre-stowed cargo, many of the provisions of Chapters 17, 18, 19 and 27 are relevant to the safe working of containers.

28.1.2   Where a container holds dangerous goods the guidance contained in section 27.2 should be followed. See also section 1.5 for guidance on the control of substances hazardous to health.

28.1.3   In handling containers, care should be taken against the possibility of uneven loading and poorly distributed or incorrectly declared weight of contents.

28.1.4   Containers carried on deck should be properly secured in such a manner as to take account of the appropriate strength features of the container and the stresses caused by the stacking of one or more upon the other.

28.1.5   Heavy items of machinery or plant and bagged bulk products which are stored on flats may need to be further secured by additional lashings.

28.1.6   On ships not specially constructed or adapted for their carriage, containers should, whenever possible, be stowed fore and aft and should be securely lashed. Containers should not be stowed over hatches unless it is known that the hatches are of adequate overall and point load-bearing strength.

28.1.7   Safe means should be provided for access to the top of a container to check lashings etc. Persons so engaged should, where appropriate, be protected from falling by use of a safety harness properly secured or by other suitable arrangements.

28.1.8   Deck lashing points which obtrude on walkways should be painted white, and warning notices displayed.

28.1.9   Adequate precautions should be taken to ensure that all electrical equipment and installations, including supply cables and connections, are safely operated and maintained.

28.1.10   Where the ship's electrical supply is used for refrigerated containers, the supply cables should be provided with proper connections for the power circuits and for earthing the container. Before use the supply cables and connections should be inspected and any defects repaired and tested by a competent person. Supply cables should only be handled when the power is switched off.

CHAPTER 29

Tankers and other ships carrying bulk liquid cargoes

29.1 General

29.1.1 Personnel joining a tanker or similar vessel for the first time should receive a basic safety induction to the main hazards associated with the vessel, either before or on joining the ship.

29.1.2 Training in emergency procedures and in the use of any special emergency equipment should be given as appropriate to members of the crew at regular intervals. The instruction should include personal first aid measures for dealing with accidental contact with harmful substances in the cargo being carried and inhalation of dangerous gases or fumes.

29.1.3 Because of the risks of ill effects arising from contamination by certain liquid cargoes, especially those carried in chemical tankers and gas carriers, personnel should maintain very high standards of personal cleanliness and particularly so when they have been engaged in cargo handling or tank cleaning. Hands should be thoroughly washed before breaks for smoking or meals. Shower baths and changes of clothing after each duty period are advisable. Dirty clothing should not be stowed in cabins.

29.1.4 Where electrical equipment is to be used in the cargo area it should be of approved design and 'certified safe'. The safety of this equipment depends on maintenance of a high order which should be carried out only by competent persons. Unauthorised personnel should not interfere with such equipment. Any faults observed, such as loose or missing fastenings or covers, severe corrosion, cracked or broken lamp glasses etc should be reported immediately.

29.2 Oil and bulk ore/oil carriers

29.2.1 Tankers and other ships carrying petroleum or petroleum products in bulk, or in ballast after carrying these cargoes, are at risk from fire or explosion arising from ignition of vapours from the cargo which may in some circumstances penetrate into any part of the ship.

29.2.2 Additionally, vapours may be toxic, some in low concentrations, and some liquid products, especially petrol (gasoline) treated with tetra-ethyl or tetra-methyl-lead, are harmful in contact with the skin.

29.2.3 Guidance on the general precautions which should be taken is given in publications of the International Chamber of Shipping:
(a) International Safety Guide for Oil Tankers and Terminals;
(b) Safety in Oil Tankers, a handbook for crew members.

Companies may additionally issue their own safety regulations. These publications should be available on board and the guidance conscientiously followed.

29.2.4   Those on board responsible for the safe loading and carriage of the cargo should have all relevant information about its nature and character before it is loaded and about the precautions which need to be observed during the voyage. The remainder of the crew should be advised of any precautions which they too should observe.

29.2.5   High risks require the strict observance of rules restricting smoking and the carriage of matches or cigarette lighters.

29.2.6   Spillages and leakages of cargo should be attended to promptly. Oil-soaked rags should not be discarded carelessly where they may be a fire hazard or possibly ignite spontaneously. Other combustible rubbish should not be allowed to accumulate.

29.2.7   Cargo handling equipment, testing instruments, automatic and other alarm systems and personal protective equipment should be maintained to a very high standard of efficiency at all times.

29.2.8   Work about the ship which might cause sparking or which involves heat should not be undertaken unless authorised after the work area has been tested and found gas-free, or its safety is otherwise assured.

29.2.9   Where any enclosed space has to be entered, the precautions given in Chapter 10 should be strictly observed. Dangerous gases may be released or leak from adjoining spaces while work is in progress and frequent testing of the atmosphere should be undertaken.

29.2.10   'Permit-to-work' procedures should be adopted unless the work to be carried out presents no undue hazard (see Chapter 7).

29.3   Liquefied gas carriers

29.3.1   Guidance on the general precautions which should be taken on these vessels is given in the Tanker Safety Guide (Liquefied Gas) published by the International Chamber of Shipping and in the publications mentioned in 29.2.3. The IMO Codes for the Construction and Equipment of Ships Carrying Liquefied Gases in Bulk contain guidance on operational aspects and the International Code (IGC Code) is statutory under the relevant Merchant shipping regulations.

MS (Gas Carriers) Regs
SI 1986 No 1073

29.3.2   The provisions in paras 29.2.4 to 29.2.10 also apply.

29.3.3   In addition, it should be noted that cargo pipes, valves and connections and any point of leakage of the gas cargo may be intensely cold. Contact may cause severe cold burns.

29.3.4   Pressure should be carefully reduced and liquid cargo drained from any part of the cargo transfer system, including discharge lines, before any opening up or disconnecting is begun.

29.3.5   Some cargoes such as ammonia have a very pungent, suffocating odour and very small quantities may cause eye irritation and disorientation together with chemical burns. Seafarers should take this into account when moving about the vessel, and especially when climbing ladders and gangways. The means of access to the vessel should be such that it can be closely supervised and is sited as far away from the manifold area as possible. Crew members should be aware of the location of eye wash and safety showers.

29.4 Chemical carriers

29.4.1 A bulk chemical tanker may be dedicated to the carriage of one or a small number of products or it may be constructed with a large number of cargo tanks in which numerous products are carried side by side simultaneously.

29.4.2 The products carried range from the so-called non-hazardous to those which are extremely flammable, toxic or corrosive or have a combination of these properties, or which possess other hazardous characteristics.

29.4.3 The ship arrangements and the equipment for cargo handling may be complex and require a high standard of maintenance and the use of special instrumentation, protective clothing and breathing apparatus for entry into enclosed spaces.

MS (IBC Code) Regs SI 1987 No 549 as amended by SI 1990 No 2602; MS (BCH Code) Regs SI 1987 No 550 as amended by SI 1990 No 2603

29.4.4 The International Maritime Organization (IMO) has produced codes (the IBC Code and the BCH Code) for the construction and equipment of ships carrying dangerous chemicals in bulk. The Codes are statutory under Merchant Shipping regulations. They contain some operational guidance, and the associated index of dangerous chemicals carried in bulk contains references to the Medical First Aid Guide for Use in Accidents Involving Dangerous Goods (MFAG) published by IMO.

29.4.5 Guidance on general operational procedures and precautions which should be followed on chemical tankers is given in the Tanker Safety Guide (Chemicals) and the booklet 'Safety in Chemical Tankers', both published by the International Chamber of Shipping. These publications, together with the codes referred to above and any special safety requirements issued by the company should be available on board.

29.4.6 Those on board responsible for the safe loading and carriage of the cargo are required to have all the relevant information about its properties and characteristics and the precautions to be observed for its safe carriage. The data should be obtained before the cargo is loaded and should be available to all concerned. The remainder of the crew should be advised of any precautions which they too should observe.

MS (Entry into dangerous Spaces) Regs SI 1988 No 1638

29.4.7 Entry into enclosed or confined spaces is subject to statutory provisions contained in Merchant Shipping regulations and the procedures and precautions given in Chapter 10 must be strictly observed.

29.4.8 'Permit-to-work' procedures should be adopted as required in Chapter 10 and described in Chapter 7 unless the work to be carried out presents no hazard.

29.4.9 Where the cargo includes packaged dangerous goods the provisions of section 27.2 apply.

29.4.10 Many products carried on chemical tankers are loosely referred to as alcohols. It should be emphasised that drinking these could lead to serious injury and death. Strict controls should be exercised when carrying such cargoes in order to prevent pilferage.

CHAPTER 30

Ships serving offshore gas and oil installations

30.1 General

30.1.1 Ships serving offshore gas and oil installations often have to operate in adverse weather conditions. Work on deck in such conditions should be avoided if the movement of the ship would create special hazards.

30.1.2 The Master of the vessel has the final responsibility for ensuring that any operation is carried out with proper regard to the safety of all those on board and that measures are taken to minimise risks.

30.1.3 Where a vessel has open stern and deck gangway doors and a low freeboard, particular care should be taken against loss of watertight integrity by ensuring that scuttles, deadlights, hatches and ventilators are securely closed. Freeing ports should be kept clear and unobstructed to ensure the rapid drainage of water trapped on the deck.

30.1.4 While work in being done on deck, the vessel should be made as sea-kindly as practicable. A look-out should be posted to give warning of imminent oncoming, quartering or following seas.

30.1.5 At all times work is being done on deck, there should be an efficient means of communication between bridge and crew. This may be provided by the use of a portable 'talk-back' speaker which can be plugged into circuit points provided at working areas.

30.1.6 During hours of darkness, sufficient lighting should be provided at access ways and at any work location, to ensure that obstructions are clearly visible and that the operation may be carried out safely.

30.1.7 Lighting should be so placed that it does not dazzle the navigational watch and does not interfere with prescribed navigation lights.

30.1.8 During bad weather, lifelines should be rigged on the working deck to facilitate safe movement. Decks should as far as practicable be kept free from ice, slush and any substance or loose material likely to cause slips or falls.

30.1.9 Men working in cold and wet conditions should wear water-proof garments over warm clothing. The need to avoid undue exhaustion and hands and limbs becoming numbed should be taken into account when making the necessary arrangements for relief at suitable intervals.

MSN No M1195 (Work Process 11) in compliance with MS (Protective Clothing and Equipment) Regs SI 1985 No 1664

30.1.10 Whenever there is a reasonably foreseeable risk of falling or being washed overboard the seaman should wear one of the forms of personal buoyancy specified in Merchant Shipping regulations and of a type which would not unduly hamper or impede working movements. If it is necessary for a man to work in an exposed position he should, where practicable, wear a safety harness and lifeline.

30.1.11 Safety helmets and high visibility garments should be worn during work on deck.

30.1.12 Advice on mooring and casting off is given in Chapter 16. Arrangements should be made to receive mooring lines, so as to avoid the necessity for seamen to jump ashore, a dangerous practice which has caused many accidents.

30.2 Carriage of cargo on deck

30.2.1 The safe securing of all deck cargoes should be checked by a competent person before the vessel proceeds on passage. To aid unloading at sea to be carried out safely, independent cargo units should, as far as practicable, be individually lashed. Lashings should, where practicable, be of a type that can be easily released and maintained (see also section 27.1).

30.2.2 All lashings should be checked at least once during each watch whilst at sea. Personnel engaged in the operation should be closely supervised from the bridge, particularly in adverse weather conditions. At night in bad weather, an Aldis lamp or searchlight should be used to aid remote checking of lashings to avoid placing men at risk.

30.2.3 Discarded rope and damaged and unserviceable equipment and cargo should not be jettisoned at sea but retained for disposal ashore. Such materials and articles can foul propellers or cause damage to fishing gear.

30.3 Lifting, hauling and towing gear

MS (Hatches and Lifting Plant) Regs SI 1988 No 1639

30.3.1 All mixed and running gear should be carefully maintained in good order and regularly inspected to detect wear, damage and corrosion. Statutory requirements for the use, maintenance and thorough examination of lifting plant are explained in Chapter 17. More frequent inspections should be made where gear has hard usage or is much exposed to sea and weather.

30.3.2 In all operations which may impose large loads or shock strains upon the gear, precautions should be taken against sudden failure which may cause injury to personnel. To the extent practicable, the system should be so designed that the weakest element is at a point where failure is likely to cause least danger.

30.3.3 While gear is under load, men essential for the operation should keep in protected positions to the greatest practicable extent. Others not engaged in the operations should keep clear of the working area.

30.4 Approaching rig and cargo handling at rig

30.4.1 Personnel should never stand forward of the windlass when letting go anchors at the rig. This is particularly important in vessels of this type because of the length of the chain and the loads thus imposed. Care should be taken when stowing the anchor cable in the locker (see Chapter 16).

30.4.2 In bad weather and under certain conditions of trim, considerable amounts of water may be shipped over the after-deck when the vessel is approaching a rig stern-on under power. Seamen should be alert to this possibility and seek positions of shelter and safety.

30.4.3   Care should be taken to keep clear when spring ropes are being lowered from the rig by crane.

30.4.4   Life-saving equipment, including lifebuoy, boathook and heaving line should be readily available at a suitable position on the stern and other points of particular danger when mooring and while cargo handling is in progress.

30.4.5   In applying the guidance of Chapter 17 to cargo handling, it should be borne in mind that the transfer of cargo at sea is at any time a difficult operation and the risks are greatly increased when heavy or bulk items are being handled from a confined deck space in a seaway.

30.4.6   When cargo is being unloaded at the rig, the lashings of each individual item or cargo should not be released until the item is about to be lifted; there are grave risks if all cargo lashings are removed before unloading operations are begun.

30.4.7   Once unlashed prior to lifting, cargo should be secured against movement as much as possible by the use of large wooden wedges, sandbags, or other effective means.

30.4.8   Personnel should be at all times alert to the danger of being hit or crushed should items of cargo swing during a lift or become dislodged through sudden movement of the ship. For this reason, all personnel should seek positions of safety as far as practicable during the lifting and lowering or cargo. If, in some circumstances, cargo hooks have to be held until the strain is taken, as when pipes are to be unloaded, crew members thus engaged should immediately move to a safe position before the actual lift is effected.

30.4.9   Lifts should be speedily effected to hoist the load well off the deck and swung clear of the ship as quickly as possible.

30.4.10   If any back-loading has to take place from the rig during off-loading of cargo from the vessel, care should be taken to ensure that the cargo taken on board is immediately secured against movement until it can be properly stowed.

30.4.11   It is essential that an efficient means of communication, preferably by radio link, is established between the rig crane operator and the working deck officer who should at all times be in visual contact with each other.

30.5   Transfer of personnel from ship to rig by 'personnel baskets'

30.5.1   The following procedures should be observed for the transference of personnel from ship to rig by 'personnel baskets':
(a) two seamen should steady the equipment when it is lowered to the deck;
(b) luggage should be secured within the net of the basket;
(c) personnel to be transferred should wear lifejackets;
(d) personnel transferring should be evenly distributed around the base board to ensure maximum stability;
(e) personnel should stand outside the basket with feet apart on the board and the basket securely gripped with both arms looped through;

(f) when the officer in charge is satisfied that all is ready, and at an appropriate moment having regard to the movement of the ship in a seaway, the basket should be lifted clear of the vessel and then swung up and out as quickly as possible before being carefully hoisted up to the rig;

(g) throughout the operation, a lifebuoy, boathook and heaving line should be kept immediately available on board the vessel for use in case of emergencies.

30.6 Transfer of personnel by boat

30.6.1 The Master of the ship providing the boat should be responsible for the operation. Due consideration should be given to the effect of prevailing conditions on the safety of the transfer.

30.6.2 The boat should be reliably powered.

30.6.3 The boat should be crewed by not less than two experienced persons, at least one of whom should be experienced in handling the boat. Lifejackets and, if necessary, suitable protective clothing, should be worn by the crew and by the personnel carried.

30.6.4 A safety rope should be provided for all personnel ascending or descending overside by ship's ladder.

30.6.5 Boarding and disembarkation should be carried out in an orderly manner under the coxswain's direction.

30.6.6 The boat's coxswain should ensure an even and safe distribution of passengers. Passengers should not stand up or change their positions during the passage between ships save under instructions from the coxswain.

30.6.7 The parent vessel should establish communication with the receiving vessel prior to the commencement of the operation and should maintain continuous visual contact with the boat concerned throughout the transfer. It is recommended that the boat should carry a VHF radio.

30.6.8 If the transfer of personnel involves a stand-by vessel, the Master should bear in mind that his vessel must at all times be ready to fulfill its stand-by vessel duties.

30.7 Anchor handling

30.7.1 Handling rig anchors at sea can be a particularly hazardous and arduous task. The vessel should be controlled in such a manner to minimise the risks concerned, in particular, to avoid as far as possible an anchor wire under heavy load whipping from quarter to quarter across the deck.

30.7.2 The provisions of section 30.3.3 on the need for crew members to keep to protected positions are particularly important during the handling of anchors and anchor buoys.

30.7.3 Anchor buoys being lifted aboard should be kept clear of the working area and lashed immediately upon landing to prevent movement.

30.7.4 Care should be taken when stoppering off wires.

30.7.5 When anchors are let go over the stern, all personnel should be well forward of the stern and in protected positions.

APPENDIX 1
British Standard specifications relevant to recommended safe working practices

NOTE: Copies of the standards produced by British Standards Institution are obtainable on microfilm or microfiche from Technical Indexes Ltd, Willoughby Road, Bracknell, Berkshire RG12 4DW, telephone 0344 426311.

A ARRANGED BY CODE CHAPTER

Chapter 5 Protective Clothing and Equipment

5.2.2	BS 5240 Industrial Safety Helmets
5.2.4	BS 4033:1966 Industrial Scalp Protectors (Light Duty)
5.4.2	BS 2092:1987 Eye-Protectors for Industrial and Non-Industrial Uses
5.5.1	BS 4275:1974 Recommendations for the Selection, Use and Maintenance of Respiratory Protective Equipment
5.5.4	BS 2091:1969 Respirators for Use against Harmful Dust, Gases and Scheduled Agricultural Chemicals BS 4555:1970 High Efficiency Dust Respirators
5.5.5	BS 4558:1970 Positive Pressure, Powered Dust Respirators
5.6.1	BS 1651:1986 Industrial Gloves BS 697:1986 Rubber Gloves for Electrical Purposes
5.6.3	BS 1870 Safety Footwear Part 1 1988 Safety Footwear Other than All-Rubber and All-Plastics Moulded Types
5.7.1	BS 1397:1979 Industrial Safety Belts, Harnesses and Safety Lanyards

Chapter 6 Signs, Notices and Colour Codes

6.2.1	BS 5378 Safety Signs and Colours Parts 1 and 2 1980 Part 3 1982
6.3.1	BS 5423:1987 Portable Fire Extinguishers
6.5.1	BS 349:1973 Identification and Contents of Industrial Gas Containers
6.5.3	BS 1319:1976 Medical Gas Cylinders, Valves and Yoke Connections

6.6.1	BS 4800:1989 Paint Colours for Building Purposes
6.6.2	BS 1710:1984 Identification of Pipelines and Services
6.7.2	BS 5609:1986 Printed Pressure-Sensitive Adhesive-Coated Labels for Marine Use, Including Requirements for Label Base Material

Chapter 8   Means of Access

8.2.3	BSMA 78:1978 Aluminium Shore Gangways
8.2.4	BSMA 89:1980 Accommodation Ladders (Obsolescent)
8.7.1	BS 3913:1982 Industrial Safety Nets

Chapter 10   Entering Enclosed or Confined Spaces

10.12.3	BS 4667 Breathing Apparatus Part 1 1974 Closed Circuit Breathing Apparatus Part 2 1974 Open Circuit Breathing Apparatus Part 3 1974 Fresh Air Hose and Compressed Air Line Breathing Apparatus Part 4 1989 Open Circuit Escape Breathing Apparatus

Chapter 12   Tools and Materials

12.3.5	BS 6500:1990 Insulated Flexible Cords and Cables

Chapter 13   Welding and Flamecutting Operations

13.2	BS 1542:1982 Equipment for Eye, Face and Neck Protection against Non-Ionising Radiation Arising during Welding and Similar Operations BS 2653:1955 Protective Clothing for Welders
13.4–13.5	BS 638 Arc Welding Power Sources, Equipment and Accessories (10 parts)

Chapter 15   Working Aloft and Outboard

15.4.2	BS 2052:1989 Ropes Made from Manila, Sisal, Hemp, Cotton and Coir BS 4128:1967 Recommendations for the Selection, Use and Care of Man-Made Fibre Ropes in Marine Applications (Historical) BS 4928:1985 Man-Made Fibre Ropes

Chapter 16   Anchoring, Making Fast, Casting off and Towing

16.2	See 15.4.2 above

Chapter 17   Lifting Plant

17.6.1 (h)	BSMA 48:1976 Code of Practice for Design and Operation of Ships' Derrick Rigs

Chapter 18   Hatches

18.4.1	BS 4263:1967 Marking of Hatchway Beams BS 4268:1967 Marking of Wooden Hatchway Covers

B BY BS NUMBER

BS NUMBER	BS TITLE	CODE REFERENCE
BS 349:1973	Identification and Contents of Industrial Gas Containers	6.5.1
BS 638	Arc Welding Power Sources, Equipment and Accessories (10 parts)	13.4–13.5
BS 697:1986	Rubber Gloves for Electrical Purposes	5.6.1
BS 1319:1976	Medical Gas Cylinders, Valves and Yoke Connections	6.5.3
BS 1397:1979	Industrial Safety Belts, Harnesses and Safety Lanyards	5.7.1
BS 1542:1982	Equipment for Eye, Face and Neck Protection against Non-Ionising Radiation Arising During Welding and Similar Operations	13.2
BS 1651:1986	Industrial Gloves	5.6.1
BS 1710:1984	Identification of Pipelines and Services	6.6.2
BS 1870	Safety Footwear	5.6.3
BS 2052:1989	Ropes Made from Manila, Sisal, Hemp, Cotton and Coir	15.4.2/16.2
BS 2091:1969	Respirators for Use Against Harmful Dust, Gases and Scheduled Agricultural Chemicals	5.5.4
BS 2092:1987	Eye-Protectors for Industrial and Non-Industrial Uses	5.4.2
BS 2653:1955	Protective Clothing for Welders	13.2.1
BS 3913:1982	Industrial Safety Nets	8.7.1
BS 4033:1966	Industrial Scalp Protectors (Light Duty)	5.2.4
BS 4128:1967	Recommendations for the Selection, Use and Care of Man-Made Fibre Ropes in Marine Applications (Historical)	15.4.2/16.2
BS 4263:1967	Marking of Hatchway Beams	18.4.1
BS 4268:1967	Marking of Wooden Hatchway Covers	18.4.1

BS NUMBER	BS TITLE	CODE REFERENCE
BS 4275:1974	Recommendations for the Selection, Use and Maintenance of Respiratory Protective Equipment	5.5.1
BS 4555:1970	High Efficiency Dust Respirators	5.5.4
BS 4558:1970	Positive Pressure, Powered Dust Respirators	5.5.5
BS 4667	Breathing Apparatus (5 parts)	10.12.3
BS 4800:1989	Paint Colours for Building Purposes	6.6.1
BS 4928:1985	Man Made Fibre Ropes	15.4.2/16.2
BS 5240	Industrial Safety Helmets	5.2.2
BS 5378	Safety Signs and Colours (3 parts)	6.2.1
BS 5423:1987	Portable Fire Extinquishers	6.3.1
BS 5609:1986	Printed Pressure-Sensitive Adhesive-Coated Labels for Marine Use, Including Requirements for Label Base Material	6.7.2
BS 6500:1990	Insulated Flexible Cords and Cables	12.3.5
BSMA 48:1976	Code of Practice for Design and Operation of Ships' Derrick Rigs	17.6.1 (h)
BSMA 78:1978	Aluminium Shore Gangways	8.2.3
BSMA 89:1980	Accommodation Ladders (Obsolescent)	8.2.4

APPENDIX 2
Bibliography

1 HMSO publications

(available from HMSO bookshops and agents, or by mail from the HMSO Publications Centre, PO Box 276, London SW8 5DT)

(a) Department of Transport Guidance and Codes of Practice
 Code of Practice for Noise Levels in Ships
 ISBN 0-11-550950-X
 Roll-on/Roll-off Ships—Stowage and Securing of Vehicles
 ISBN 0-11-550995-X
 Ship Captain's Medical Guide
 ISBN 0-11-550684-5

(b) Health and Safety Executive Guidance and Codes of Practice
 Anthrax: health hazards (Guidance note EH 23)
 ISBN 0-11-883194-1
 Asbestos (Guidance Note MS 13)
 ISBN 0-11-885402-X
 Beryllium—Health and Safety Precautions (Guidance Note EH 13)
 ISBN 0-11-883038-4
 Colour Vision (Guidance Note MS 7)
 ISBN 0-11-883950-0
 Control of Substances Hazardous to Health: Control of Carcinogenic Substances. Control of Substances Hazardous to Health Regulations 1988. Approved Code of Practice
 ISBN 0-11-885468-2
 Drilling Machines: Guarding of Spindles and Attachments (Guidance Booklet HS(G)44)
 ISBN 0-11-885466-6
 Entry into Confined Spaces (Guidance Note GS 5)
 ISBN 0-11-883067-8
 Fumigation using Methyl Bromide (Bromomethane) (Guidance Note CS 12)
 ISBN 0-11-883549-1
 A Guide to Dangerous Substances in Harbour Areas Regulations, 1987
 ISBN 0-11-883991-8
 Hazard and Risk Explained. Control of Substances Hazardous to Health Regulations 1988 (COSHH). Health and Safety Information and COSHH (Leaflet IND(G)67(L))
 Industrial Use of Flammable Gas Detectors (Guidance Note CS 1)
 ISBN 0-11-883948-9
 Introducing Assessment (Leaflet INF(G)64(L))
 Occupational Exposure Limits (Guidance Note EH 40)
 ISBN 0-11-885420-8
 Protection against Electric Shock (Guidance Note GS 27)
 ISBN 0-11-883583-1

Respiratory Protective Equipment: a Practical Guide for Users (Guidance Booklet HS(G)53)

ISBN 0-11-885522-0

Respiratory Potective Equipment for Use against Asbestos (Guidance Note EH 41)

ISBN 0-11-883512-2

Safety in Docks: Docks Regulations 1988: Approved Code of Practice with Regulations and Guidance

ISBN 0-11-885456-9

Safety in the Use of Abrasive Wheels (Guidance Booklet HS(G)17)

ISBN 0-11-883739-7

Visual Display Units

ISBN 0-11-883685-4

Note: Further information and a comprehensive list of HSE guidance is available from the HSE Library and Information Service at Baynards House, 1 Chepstow Place, Westbourne Grove, London W2 4TF, telephone 071-221 0870

(c) Statutory Instruments (SIs)

To purchase a copy of an Act of Parliament (eg a Merchant Shipping Act) or a Regulation made under such an Act from HMSO you should quote the number of the relevant SI. The SI numbers of some of the main regulations relating to safe working practices on board ship are included in Appendix 3.

2 Department of Transport free publications

(a) Booklets
(available from Industrial Services Consortium, The Paddock, Frizing Hall, Bradford BD9 4HD, telephone 0274 541391 or (for single copies) the Department of Transport Marine Library, Sunley House, 90 High Holborn, London WC1V 6LP).

Fire on Ships

Guidance Notes to the Merchant Shipping (Health and Safety: General Duties) Regulations 1984

Personal Survival At Sea

(b) Merchant Shipping Notices (MSN)
About 40 new MSNs are issued each year and any printed list quickly becomes out of date, but a current list is always available from the Department of Transport Marine Library.

3 IMO publications

(available from the International Maritime Organization, Publications Section, 4 Albert Embankment, London SE1 7SR, telephone 071-735 7611).

Code for the construction and equipment of ships carrying dangerous chemicals in bulk (BCH Code) (1990 edn)
IMO sales no: IMO-772E   ISBN 92-801-1255-4

Code for the construction and equipment of ships carrying liquefied gases in bulk (1983 edn)
IMO sales no: IMO-782E   ISBN 92-801-1165-5

Code of safe practice for solid bulk cargoes (1989 edn)
IMO sales no: IMO-260E   ISBN 92-801-1245-7

Code for existing ships carrying liquefied gases in bulk (1976 edition)
IMO sales no: IMO-788E ISBN 92-801-1051-9

Emergency procedures for ships carrying dangerous goods: see Supplement to IMDG Code

International code for the construction and equipment of ships carrying dangerous chemicals in bulk (IBC Code) (1990 edition)
IMO sales no: IMO-100E ISBN 92-801-1254-6

International code for the construction and equipment of ships carrying liquefied gases in bulk (IGC Code) (1983 edition)
IMO sales no: IMO-104E ISBN 92-801-1163-9

International maritime dangerous goods code (IMDG Code)
(1990 edition comprises 4 loose-leaf volumes)
IMO sales no: IMO-200E ISBN 92-801-1243-0

International maritime dangerous goods code: Supplement
(The contents of the 1990 Supplement are:
- Emergency procedures (EmS)
- Medical first aid Guide (MFAG)
- Solid chemicals in bulk (BC Code)
- Reporting procedures
- Packing cargo transport units
- Use of pesticides in ships)

IMO sales no: IMO-210-E ISBN 92-801-1248-1

Medical First Aid Guide for Use in Accidents Involving Dangerous Goods (MFAG): see Supplement to IMDG Code.

4 International Chamber of Shipping publications

(obtainable from Witherby and Co Ltd, 32-36 Aylesbury Street, London EC1R 0ET, telephone 071-251-5341)

International safety guide for oil tankers and terminals (3rd edition, 1988)
ISBN 0-948691-62-X

Safety in chemical tankers (1977 booklet)
Safety in liquefied gas carriers (1980 booklet)
Safety in oil tankers (1973 booklet)
Tanker safety quide (chemicals)
(In preparation)
ISBN 0-948691-50-6

Tankers safety guide (liquefied gas) (1978)
(New edition in preparation)
ISBN 0-906270-01-4

APPENDIX 3
Merchant shipping health and safety law

If you work on a ship registered in the United Kingdom your health and safety at work are protected by law. Here is a brief guide to health and safety law for seafarers. It does not describe the law in detail but it does list the key points for seafarers working on United Kingdom ships (other than fishing vessels) on which merchant shipping laws apply.

Your employer has a duty under the law to ensure, so far as is reasonably practicable, your health and safety and that of other persons aboard ship who may be affected by what he does or does not do.

In general your employer's duties include:—

Merchant Shipping (Health and Safety: General Duties) Regulations SI 1984: No. 408

—ensuring plant and machinery are safe and safe systems of work are set and followed;

—ensuring articles and substances are moved, stored and used safely;

—giving you the information, instruction, training and supervision necessary for your health and safety;

—keeping the working areas on your ship safe and without risks to health;

—maintaining a safe working environment on board.

Your employer must also:

—draw up and bring to your attention a statement setting out his health and safety policy and explaining how he will carry it out (but only if he employs at least five people on United Kingdom ships);

MS (Protective Clothing and Equipment) Regulations SI 1985: No. 1664

—provide any protective clothing or equipment specifically required by merchant shipping law.

The owner of your ship (who may also be your employer) has other duties—for example:

MS (Code of Safe Working Practices) Regulations SI 1980: No. 686

—making sure your ship carries the specified number of copies of the Code of Safe Working Practices; whilst your Master must make sure you have access to a copy.

Your Master and his employer (who may also be your employer) must make sure that:

MS (Means of Access) Regulations SI 1988: No. 1637 and MS (Safe Movement on Board Ship) Regulations SI 1988: No. 1641

—there is a safe means of access to the ship and to any place on the ship to which you may be expected to go;

MS (Entry into Dangerous Spaces) Regulations SI 1988: No. 1638

—procedures for safe entry and working in dangerous spaces are clearly laid down and observed;

MS (Guarding of Machinery and Safety of Electrical Equipment) Regulations SI 1988: No. 1636

—dangerous parts of the ship's machinery are securely guarded or kept as safe to anyone on board as if they were securely guarded;

—all ship's electrical equipment and installations are so constructed, installed, operated and maintained that the ship and all persons are protected against electrical hazards;

MS (Hatches and Lifting Plant) Regulations SI 1988: No. 1639

—all hatch coverings are sound, strong and properly maintained; they are used only if they can be removed and replaced without endangering anyone; and their correct replacement position is clear;

—all ship's lifting plant (hoists, lifts, chains, ropes, cranes, lifting tackle, etc) is well constructed, well maintained and examined at specified intervals, and is used only by competent authorised operators.

<small>MS (Safety Officials and Reporting of Accidents and Dangerous Occurrences) Regulations SI 1982: No. 876 (the "SORADO" Regulations): Part 1</small>

If your ship goes to sea with more than five crew the employer must appoint a safety officer whom he and the Master must consult on matters affecting your health and safety. You may also elect your own official safety representative. If you do the employer and the Master must consult him too.

<small>MS ("SORADO") Regulations: Part 2</small>

Your Master (or the next most senior officer available) must report certain accidents and dangerous occurrences to the Department of Transport.

As an employee, you have legal duties too. They include:—

<small>MS (Health and Safety: General Duties) Regulations SI 1984: No. 408, Regulations 5 and 7</small>

—taking reasonable care for your own health and safety and that of others aboard ship who may be affected by what you do or do not do;

—co-operating with your employer on health and safety, or with anyone else who is responsible for it;

—not interfering with or misusing anything provided for your health or safety.

There are other rules that are also meant to protect your health and safety. There are regulations for the construction and equipment of ships, life-saving appliances, fire protection, the prevention of collisions, load line matters, dangerous goods, crew accommodation, medical stores, provisions and water etc. These are all listed in a Merchant Shipping Notice which is issued by the Department of Transport and regularly revised and reissued as changes in the law are introduced.

Merchant Shipping Notices are also used by the Department of Transport to bring safety and other matters to your attention and copies of the current ones should be on board your ship.

The main Acts of Parliament governing health and safety on ships are the Merchant Shipping Acts but for particular purposes other Acts (and Regulations made under these) may be relevant. For example the Health and Safety at Work, etc Act 1974 applies to shore based workers (eg dockers and ship repairers) working on ships, and to certain small inland waterway or harbour craft not covered by Merchant Shipping Regulations. The provisions of this Act give similar protection and impose similar duties to those set out here.

International laws and agreements, and conventions or directives produced by such bodies as the International Labour Organisation (ILO), the International Maritime Organisation (IMO) and the European Community all help decide the shape of our national laws.

If you think there is a health or safety problem on your ship you should first discuss it with your safety officer or safety representative, or with your employer. If the problem remains or you need more help, surveyors from a Department of Transport Marine Office can give advice on how to comply with the law.

APPENDIX 4
Lifting plant (Chapter 17)

Contents of Appendix 4:—

A Examples of hand signals recommended for use with lifting appliances on ships (see paragraph 17.2.28).
(NOTE: the examples shown have been in common use in United Kingdom registered ships in recent years; but other standards may also be encountered).

B Certificates of test and thorough examination (see paragraph 17.15.4):

    1 Lifting appliances

    2 Derricks used in union purchase

    3 Loose gear

    4 Wire rope.

C Register of ships' lifting appliances and cargo handling gear (see paragraph 17.15.5).

APPENDIX 4A
Code of Hand Signals

APPENDIX 4B

CERTIFICATE OF TEST AND THOROUGH EXAMINATION OF LIFTING APPLIANCES

Name of Ship

Certificate No. ....

Official Number

Call Sign

Port of Registry

Name of Owner

(1) Situation and description of lifting appliances (with distinguishing numbers or marks, if any) which have been tested and thoroughly examined	(2) Angle to the horizontal or radius at which test load applied	(3) Test load (tonnes)	(4) Safe working load (SWL) at angle or radius shown in column 2 (tonnes)

Name and address of the firm or competent person who witnessed testing and carried out thorough examination

....
....
....

I certify that on the date to which I have appended my signature, the gear shown in column (1) was tested and thoroughly examined and no defects or permanent deformation were found; and that the safe working load is as shown.

Date: .... Signature: ....

Place: ....

Note: This Certificate is the standard international form as recommended by the International Labour Office in accordance with ILO Convention No. 152.

APPENDIX 4B *continued*

CERTIFICATE OF TEST AND THOROUGH EXAMINATION OF DERRICKS USED IN UNION PURCHASE

Name of Ship   Certificate No. ....

Official Number

Call Sign

Port of Registry

Name of Owner

(1)	(2)	(3)	(4)
Situation and description of derricks used in union purchase (with distinguishing numbers or marks) which have been tested and thoroughly examined	Maximum height of triangle plate above hatch coaming (m) or maximum angle between runners	Test load (tonnes)	Safe working load, SWL (U), when operating in union purchase (tonnes)

Position of outboard preventer guy attachments: (a) forward/aft* of mast .... (m)
and
(b) from ship's centre line .... (m)

Position of inboard preventer guy attachments: (a) forward/aft* of mast .... (m)
and
(b) from ship's centre line .... (m)

* Delete as appropriate

Name and address of the firm of competent person who witnessed testing and carried out thorough examination

....
....
....

I certify that on the date to which I have appended my signature, the gear shown in column (1) was tested and thoroughly examined and no defects or permanent deformation were found; and that the safe working load is as shown.

Date: ....   Signature: ....

Place: ....

Note: This Certificate is the standard international form as recommended by the International Labour Office in accordance with ILO Convention No. 152.

APPENDIX 4B *continued*

CERTIFICATE OF TEST AND THOROUGH EXAMINATION OF LOOSE GEAR

Name of Ship

Certificate No. ....

Official Number

Call Sign

Port of Registry

Name of Owner

Distinguishing number or mark	Description of loose gear	Number tested	Date of test	Test loaded (tonnes)	Safe working load (SWL) (tonnes)

Name and address of makers or suppliers: ....

Name and address of the firm or competent person who witnessed testing and carried out thorough examination

....
....
....

I certify that the above items of loose gear were tested and thoroughly examined and no defects affecting their SWL were found.

Date: .... Signature: ....

Place: ....

Note: This Certificate is the standard international form as recommended by the International Labour Office in accordance with ILO Convention No. 152.

APPENDIX 4B *continued*

CERTIFICATE OF TEST AND THOROUGH EXAMINATION OF WIRE ROPE

Name of Ship   Certificate No. ....

Official Number

Call Sign

Port of Registry

Name of Owner

Name and address of maker or supplier

Nominal diameter of rope (mm)

Number of strands

Number of wires per strand

Core

Lay

Quality of wire (N/mm^2)

Date of test of sample

Load at which sample broke (tonnes)

Safe working load of rope (tonnes)

Intended use

Name and address of the firm of competent person who witnessed testing and carried out thorough examination

....
....
....

I certify that the above particulars are correct, and that the rope was tested and thoroughly examined and no defects affecting its SWL were found.

    Date: ....   Signature: ....

    Place: ....

Note: This Certificate is the standard international form as recommended by the International Labour Office in accordance with ILO Convention No. 152.

APPENDIX 4C

*REGISTER OF SHIPS' LIFTING APPLIANCES AND
CARGO HANDLING GEAR*

Name of Ship

Official Number

Call Sign

Port of Registry

Name of Owner

Register Number

Date of Issue

Issued by

Signature and Stamp

Note: This Register is the standard international form as recommended by the International Labour Office in accordance with ILO Convention No. 152.

APPENDIX 4C *continued*

PART I—Thorough examination of lifting appliances and loose gear

(1) Situation and description of lifting appliances and loose gear (with distinguishing numbers or marks, if any) which have been thoroughly examined (see Note 1)	(2) Certificate Nos.	(3) Examination performed (see Note 2)	(4) I certify that on the date to which I have appended my signature, the gear shown in column (1) was thoroughly examined and no defects affecting its safe working condition were found other than those shown in column (5) (Date and signature)	(5) Remarks (To be dated and signed)

Note 1: If all the lifting appliances are thoroughly examined on the same date it will be sufficient to enter in column (1) "All lifting appliances and loose gear". If not, the parts which have been thoroughly examined on the dates stated must be clearly indicated.

Note 2: The thorough examinations to be indicated in column (3) include:
    (a) Initial.
    (b) 12 monthly.
    (c) Five yearly.
    (d) Repair/damage.
    (e) Other thorough examinations including those associated with heat treatment.

APPENDIX 4C *continued*

PART II—Regular inspections of loose gear

(1) Situation and description of loose gear (with distinguishing number or mark, if any) which has been inspected (see Note 1)	(2) Signature and date of the responsible person carrying out the inspection	(3) Remarks (To be dated and signed)

Note 1: All loose gear should be inspected before use. However, entries need only be made when the inspection discloses a defect.

INDEX

Abandon ship
— drills 3.1.4

Abrasions 1.2.4; 23.3.3-4; 25.1.3-4

Abrasive wheels 12.5

Access
— to hold 9.6.4-5
— to ship 8

Access equipment 8.1
— construction 8.2
— corrosion 8.3.4
— guard ropes 8.4.4
— inspection 8.3.2; 8.4.4
— maintenance 8.3
— positioning 8.4
(See also Gangways; Ladders)

Accidents 1.1.14; 4.10
— catering staff 25.2
— injured persons 4.10.4
— investigation 4.10
— laundries 26.2
— manual lifting 11.1.1; 11.2.14
— record book 4.6.6; 4.10
— slips and falls 5.7

Accommodation ladders 8.2.1-2; 8.2.4
— angle of 8.4

Acetylene
— welding 13.6

Acid
— in batteries 24.2
— in domestic cleaning substances 25.1.12

Aerosols 1.4.12

Air pockets
— in hydraulic systems 21.1.5

Alarm systems
— isolation of 22.1.3

Alcohol 1.2.17
— cargo 29.4.10
— misuse 1.2.17

Aloft 15
— painting 14.3
— work 15

Aluminium
— corrosion 8.10

Ammonia
— as bulk cargo 29.3.5

Anchors 16.1

Anchor handling
— offshore supply ships 30.4.1
— rig anchors 30.7

Anchoring 16.1

Anthrax 1.2.9

Asbestos
— and smoking 1.2.19
— danger from dust 1.6 (see also 1.5)
— emergency repairs 1.6.4
— removal in port 1.6.3

Asbestos lagging
— removal 1.6.4; 22.2.4

Atmosphere
— enclosed space 10

Auxiliary machinery 22.7

Back injuries 11.1.1

Backfiring
— of gas cylinders 13.6.6

Backlash
— man-made fibre ropes 16.2.8

Batteries 24
— alkaline 24.3; 24.1.17
— general precautions 24.1
— lead-acid 24.2; 24.1.17

BCH Code 29.4.4; Appendix 2 (under 'IMO publications')

Beams
— use with hatchways 18.4

Bench machines 12.4

Beryllia 23.3.3–5

Bights of rope
    — safety hazard 16.3.11; 16.5.2; 16.6.7

Blankets
    — combustion risk 2.4.2

Bleach 1.2.13; 12.9.3; 25.1.12

Blocks
  see Lifting gear

Boarding ladders 16.4.2

Body protection 5.8

Boilers 20.2; 22.6
    — avoidance of blowbacks 20.2.2
    — display of operating instructions 20.2.1
    — overhaul 22.6

Boiler suits
    — for welding 13.2.2

Bosun's chair 15.3

Brake operators
    — head and eye protection 16.1.2

Brass turning
    — eye and face protection 12.4.11–13

Breathing apparatus 5.5.9–11
    — checking 10.12.6
    — compressed air 10.12.2–4
    — use in dangerous spaces 10.8.8; 10.10.2–5; 10.12
    — use in fire drills 3.2.7
    — use in hot work 13.1.4
    — use when paint spraying 14.2.5

Bulk liquid cargoes 29
    — chemicals 29.4
    — general precautions 29.1
    — liquefied gas 29.3
    — oil and bulk ore/oil carriers 29.2

Bulkheads
    — hot work alongside 13.3.5

Bulwark ladders 8.2.5

Bump caps 5.2.4

Buoyancy aids
    — for crew on offshore supply ships 30.1.10

Buoyancy garments
    — use when working overside 14.3.4; 15.1.3

Buoys
    — mooring to 16.4

Burns
— hazard in laundries 26.2

Cables—see electrical cables and Leads

Can openers 25.5.4

Car decks
— retractable 17.2.7–8

Carbolic acid 25.1.12

Carbon monoxide 10.4.15

Carbon tetrachloride 22.9.5; 23.1.5

Cargo
— bulk liquid 29
— chemicals 29.4
— dangerous goods 27.2
— liquefied gas 29.3
— loading of solid bulk cargoes 27.2.4
— oil 29.2
— on deck 9.2.9
— on offshore supply ships 30.2
— stowage 27.1
— transfer at sea 30.4.5 et seq.
— working 27.3

Cargo ships 27
— dangerous goods 27.2
— general cargo 27
— liquids 29
see also 11, 17, 18 and 19

Cargo spaces 19
— access 19.1
— fencing 19.3
— general precautions 19.4
— lighting 19.2
— removal of injured persons 19.1.3
— securing of cargo stacks 19.4.2–4

Carrying 11

Casting off 16.5

Catering equipment 25.4

Cathode ray tubes 23.3.2 et seq.

Caustic soda 1.2.13; 12.9.3; 25.1.12

Certification of lifting plant 17.15
— approved forms 17.15.4; Appendix 4B

Chain stoppers 17.8.1–2

Chemical agents 12.9
— mixing of 12.9.4
— unlabelled 12.9.1
— skin and eye protection 12.9.2

Chemical tanks 10.4.13-19; 10.4.21-23

Chisels 12.2.3-7
— vibration from 12.3.14

Choppers
— for catering 25.5

Chuck keys 12.4.14

Cigarettes
— fire risk 2.1

Cleaning agents 12.9.3

Clothing 1.3
— protective 5
(See also Protective Clothing and Equipment)

Code of Safe Working Practices 1.1.1-8
— application 1.1.4-7
— copies to be carried on board 1.1.8
— Regulations 1.1.8

Codes of practice and guides
— bulk chemicals 29.4.4-5
— bulk liquefied gases 29.3.1
— dangerous goods (IMDG Code) 27.2.1-2; 27.2.4; (emergency procedures) 27.2.11
— medical first aid (MFAG) 27.2.5 (for use in accidents with dangerous goods)
— oil tankers 29.2.3
— solid bulk cargoes 27.2.4
— stowage and securing of vehicles on ro/ro ships 8.9.4
(See also Appendix 2)

Cold chisels 12.2.3; 12.2.7

Cold stores 25.6

Colour coding 6
— dangerous goods 6.7.1-2
— electrical wires and cables 6.4.1-2
— gas cylinders 6.5
— pipelines 6.6

Competency
— certificates of 17.2.10

Competent person
— assessment of dangerous atmosphere 10.3.1-2
— certification of lifting plant 17.15.2
— examination of lifting plant 17.13

Compressed air
— operation of compressed air tools 12.7

Compressed gas cylinders
— colour coding 6.5
— general guidance 12.8.1–5
— storage 12.8.3
— valves 12.8.3

Confined spaces 10
see also Dangerous spaces

Container terminals
— access hazards 8.9.5

Containers 28
— access 28.1.7
— carriage 28
— stowage 28.1.6

Control of substances hazardous to health 1.5

Cooking appliances 25.3

Cooking oil and fat 25.3.8–14
— fire prevention 2.6.1–2

Corrosion
— of aluminium ladders etc 8.3.4

Coughs 1.2.19

Cradles 15.2
— ancillary equipment 15.2.4–7
— construction 15.2.1–2
— rigging 15.2.5

Cranes 17.5
— crane drivers 27.3.4–5
— safe use 17.3; 17.5

Cuts and abrasions 1.2.4; 23.3.3–4; 25.1.3–4

Cylinders
— compressed gas 12.8
see also Gas cylinders

Dangerous goods 27.2
— handling 27.2.2–8
— hazard diagrams 6.7.2
— IMDG Code 27.2.1–2
— labelling 6.7.2
— leakage and spillage 27.2.9–11
— risk assessment and control 1.5

Dangerous Goods and Marine Pollutants Regulations 6.7.1; 27.2.1

Dangerous occurrences
— recording of 4.6.6; 4.10.13–14

Dangerous spaces 10
- chemical tanks 10.4.13-19; 10.4.21-23
- drills and rescue 10.11
- duties of employer 10.15.1
- entry 10.8-10
- maintenance of equipment 10.13
- oil tanks 10.4.10-12
- permit to work system 10.7
- precautions 10.7
- preparations for entry 10.5
- rescue equipment 10.8.3; 10.8.8-9; 10.10.6
- testing of atmosphere 10.6; 10.9.2
- training and instruction 10.14
- use of breathing apparatus 10.8.3; 10.10; 10.12

Deck
- drainage 9.8.1
- duck boards 9.8.4
- gratings 9.2.3
- lighting of openings 9.3.7
- securing of cargo 9.2.9
- walkways 9.2.1

Deep fat fryers 25.3

Dermatitis 1.2.7-8; 1.1.12

Derricks 17.6-7
- rigging 17.6.1
- safe use 17.6.2 et seq.
- union purchase 17.7

Diarrhoea 25.1.6

Diesel engines 22.7.2-6

Diseases 1.2; 25.1.5-25.1.6

Dockside
- access routes 8.9.5

Doors
- fire doors 3.2.8
- securing 1.4.7
- watertight 9.9

Drilling tools 12
- fixed installations 12.4
- pneumatic drills 12.3
- hand tools 12.2

Drills and rescue 3
- abandon ship drills 3.1.4
- fire drills 3.2
- rescue from dangerous space 10.11

Drivers
- of ships' vehicles 9.7.1

Drowning
— protective equipment 5.9

Drugs 1.2.17

Drying
— of clothing 2.2.8
— cabinets 2.2.9

Dry-cleaning 26.4
— hazards of solvent vapour 26.4.1–3
— responsible officer 26.4.5–6

Duck boards 9.8.4

Dust masks 5.5.4; 14.1.2

Dust respirators 5.5.4
— positive pressure 5.5.5

Dysentery 25.1.6

Ear protection 5.3; 20.1.3; 22.2.2

Electric heaters
— guards required 2.2.8

Electric shock
— posting of first aid instructions 22.9.2
— power tools 12.3.2–3
— welding operations 13.5.1–7

Electric welding 13.4–5
— assistants 13.5.2
— cables 13.4.3–7
— electric shock 13.5.1–7
— power sources 13.4.1–2; 13.4.10–12
— precautions 13.5
— protective clothing 13.2.2; 13.5.1

Electrical appliances
— fire risk 2.2
— faults to be reported 2.2.2
— isolation during overhaul 22.1.5; 22.7.1; 22.9.3
— portable appliances isolated after use 2.2.7
— to be firmly secured 2.2.3

Electrical cables and leads
— on power tools 12.3.4
— use with refrigerated containers 28.1.10
— welding operations 13.4.3–7

Electrical equipment
— in dangerous spaces 10.8.2
— isolation during overhaul 22.1.5; 22.7.1; 22.9.3
— overhaul 22.9
— work on live equipment 22.9.6 et seq.
see also Electrical appliances etc.

Electrical plugs
— domestic 6.4.2
— makeshift plugs not to be used 2.2.5

Electrical wiring 1.4.4; 6.4
— on power tools 12.3.4-7
— overloading 2.2.6

Electrolytic capacitors 23.2.8

Electronic equipment 23
— electrical hazards 23.2

Emergency escape routes
— access to 22.1.13

Employers
— duties in respect of dangerous spaces 10.15.1
— duties towards safety officials 4.9
— duty to ensure safe access 8.1.1
— general duties 1.1.9
— responsibilities in respect of lifting plant 17.2.1; 17.2.5-6
— responsibility for safe movement on board 9.1.1
— responsibility for safety in manual lifting operations 11.1
— responsibility for safety of work equipment 12.1.1

Enclosed spaces 10
— general guidance 10.1
— permit to work system 7.1.1-2; 7.1.5
— precautions 10.2
see also Dangerous spaces

Engine room
— bilges 20.1.9
— fire pumps 3.2.1; 20.1.10

Engines
— overhaul of 22.8

Entry into Dangerous Spaces Regulations 10.15.1; 29.4.7

Equipment 12
— reporting of defects 9.10.2

Eamination of lifting plant 17.13
— certificates and reports 17.15

Exhaust pipes 20.1.2

Explosions
— hazard in battery compartments 24.1
— hazard from petroleum in bulk 29.2.1
— precautions during hot work 13.3

Explosives
— as cargo 27.2.7

Eye protection
- brake operators 16.1.2
- from abrasive wheels 12.5.4
- from chemical agents 12.9.2
- from wirebrushing, brass turning etc 12.4.11
- paint sprayers 14.2.2-3
- when overhauling machinery 22.2.3

Eyes
- protection 5.4

Eye-bolts 22.3.2

Face
- protection 5.4

Falls
- protection from 5.7

Fat
- for cooking 25.3.8-14

Fatigue
- from lifting and carrying 11.2.14-15

Feet
- protection 5.6

Fencing
- on openings and hatchways 9.5.3-4

Ferries
- pedestrian ramps 8.9.4

Fire
- hazard from petroleum in bulk 29.2
- precautions during hot work 13.3

Fire appliances
- inspection 3.2.8

Fire drills 3.1.4; 3.2

Fire extinguishers 3.2.5-6; 3.4.5; 6.3
- for hot work 13.3.7

Fire fighting 3.4
- escape from smoke 3.4.7
- raising alarm 3.4.4
- reduction of air supply 3.4.6

Fire fighting equipment
- access to 22.1.13
- during loading of explosives or flammables 27.2.7

Fire main
- interruption of water supply 22.1.2

Fire precautions 2

Flamecutting 13
- fire and explosions 13.3
- fumes 13.1.4
- general guidance 13.1; 13.6
- inspection and repair of equipment 13.1.5
- protective clothing 13.2
- supervision 13.3.7
- testing for gases in tanks and holds 13.3.6
- ventilation 13.1.4

Flammability
- testing 10.6.8-11; 10.6.14
- vapours in enclosed spaces 10.4.19-20

Flammable liquids 27.2.7

Floor plates 22.2.5; 22.4

Food 25
- cooking appliances 25.3
- health and hygiene 25.1
see also Galleys

Footwear 1.3.5; 5.6.2-3
- for use in galley 25.2.1
- slip resistant, for use when overhauling engine room machinery 22.2.5

Fumes
- toxic 10.1.3
- welding and flamecutting operations 13.1.4

Funnel
- safety precautions when working on 15.16

Galleys
- catering equipment 25.4
- cooking appliances 25.3
- drainage 9.8
- fire prevention 2.6; 25.3.1; 25.3.8-13
- health and hygiene 25.1
- knives, saws, choppers etc 25.5
- refrigerated rooms and store rooms 25.6
- slips, falls and tripping hazards 25.2
- working clothes 1.3.2

Gangways 8.2.1-3
- aluminium 8.10
- angle 8.4
- examination 8.10.4

Garbage
- in galley 25.1.13
(See also Rubbish)

Gas
- combustible 10.4.19-20; 10.6.8-11; 10.6.14
- toxic 10.1.3; 10.6.12-14

177

Gas cylinders 6.5; 13.6
— backfiring 13.6.6
— medical 6.5.3
— overheating 13.6.6
(See also Compressed gas cylinders)

Gas monitoring equipment 10.6; 10.13.2

Gas welding 13.6
— backfires 13.6.6
— blowpipes 13.6.10–18
— flashbacks 13.6.11
— manifold hose connections 13.6.9
— overheating of cylinders 13.6.6

Gasoline 29.2.2

General Duties Regulations 1.1.9

Gloves 1.3.6; 5.6.1
— for hot work 13.2.2
— use in dry cleaning 26.4.4

Grain cargo
— toxic gases 10.4.17

Gratings 9.2.3

Guard rails
— cargo spaces 19.3

Guarding of machinery 20.1.1 et seq.
— replacement of guards after repairs 22.1.16

Guards
— on abrasive wheels 12.5.9–10
— on catering equipment 25.4.2–3; 25.4.5
— on heaters 2.2.8; 2.2.10
— on laundry equipment 26.3.2
— on machinery 12.4.2–3
— on openings 9.5.1

Hair
— protection 5.2.5

Hammers
— hand held 12.2
— power 12.3
— vibration from 12.3.14

Hand lamps
— voltage 12.3.2

Hand signals 17.2.28; Appendix 4A

Hand tools 12.2
— manner of use 12.2.5–8

Hands
- protection 5.6
- washing before food handling 25.1.2

Hatch covers 18.2-4
- inspection 18.2.1
- instruction of operators 18.2.6
- maintenance and repair 18.2.2
- manually handled 18.4.3-4
- mechanical 18.3
- non mechanical covers and beams 18.4
- replacement of covering 18.2.7
- steel hinged inspection and access lids 18.5

Hatches 18
- safety during interruptions in loading 27.3.24

Hatches and Lifting Plant Regulations
- hatches 18
- lifting plant 17

Hatchways
- access 9.6.4-5
- closing of 9.5.2
- guarding of 9.5.1

Hazards
- control 1.1.12; 1.5.8; 7.1.2
- elimination 7.1.2
- removal 1.1.11
- reporting of 9.10.2

Hazard diagrams 6.7.2

Hazard warnings 9.4.2

Head
- protection 5.2

Head protection
- brake operators 16.1.2
- towing operations 16.6.6

Health 1.2
- cuts and abrasions 1.2.4; 23.3.3-4; 25.1.3-4
- diseases 1.2; 25.1.5-6
- heat exhaustion and stroke 1.2.14
- hygiene 1.2.1; 25.1; 29.1.3
- infections 1.2.3-5; 25.1.5
- skin diseases 1.2.7-13; 26.4.4
- vaccination 1.2.5
(See also Anthrax; Asbestos; Drugs; Malaria; Salt tablets; Smoking; Vibration "white finger" etc)

Hearing protectors 5.3
- use in machinery spaces 20.1.3; 22.2.2

Heat stroke 1.2.14

Heaters
— construction of 2.2.12
— electrical 2.2.8
— portable 2.2.10–11

Helmets 5.2

High visibility
— on offshore supply vessels 30.1.11

Hoses
— fires and fire drills 3.2.3
— for gas welding and flamecutting 13.6.9–14
— safe use when paint spraying 14.2.9–10

Hot work 13
— permit to work system 7.1.5
see also Flamecutting and Welding

Hydraulic equipment 21
— general precautions 21.1
— hydraulic fluid 21.1.6–7; 22.12.5; 22.12.7
— hydraulic jacks 21.2
— repairs and maintenance 22.12

Hydrocarbons
— monitoring 10.6.5
— toxicity 10.4.10–12

Hydrogen sulphide 10.4.17

Hygiene 1.2.1
— in galleys 25.1
— on tankers 29.1.3

IBC Code 29.4.4; Appendix 2 (under 'IMO publications')

IMDG Code 1.5.6; 27.2.1–2; 27.2.4; Appendix 2 (under 'IMO publications')
— packaging and labelling of dangerous goods 6.7.2

Injured persons
— priority over accident investigation 4.10.4
— removal from hold 19.1.3
— rescue from dangerous space 10.11.5

Inspections
— access equipment 8.3.2
— check-list for safety officers 4.6.17–21

Insulation
— heated surfaces 20.1.2
— of tools 12.3

Jacks
— hydraulic 21.2

Knives
— catering 25.5

Ladders
- accommodation 8.2.1–4
- aluminium 8.10
- angle of 8.6.2
- bulwark 8.2.5
- portable 8.6; 9.6.6–7; 15.5
- rope 8.2.6; 8.6
- to holds 9.6.4–5

Lamps
- spirit 12.6
see also Lighting

Lathes 12.4
- fencing of projecting material 12.4.18
- removal of chuck key 12.4.14

Laundry
- burns and scalds 26.2
- drainage of area 9.8.1
- dry cleaning 26.4
- fire prevention 26.5
- fire risk 2.3.1
- general precautions 26.1
- machinery and equipment 26.3

Lead
- from paints 14.1.2; 14.2.4

Leads
- on portable or temporary lights 9.3.8–9

Lifebuoys
- at access point 8.8.1
- use when working overside 14.3.4

Lifejackets
- transfer of crew by boat 30.6
- transfer of crew by personnel basket 30.5
- use during drills 3.1.5

Lifelines 5.7; 9.2.7
- on offshore supply vessels 30.4.4
- use when painting overside 14.3.4

Liferafts
- davit-launched 3.3.12
- drills 3.3
- instructions 3.3.20

Lift trucks 17.2.47
see also Trucks

Lifting 11
- posture 11.2.3–4

Lifting gear 17.2.1–2; 17.2.36–39; 17.2.41; 17.9
- overhaul of 17.9

Lifting plant 17
- certificates and reports 17.15
- construction 17.2.1
- controls 17.2.15–20
- cranes 17.5
- defects 17.11
- examination 17.13
- maintenance 17.2.1; 17.2.3
- marking 17.2.41; 17.14
- on offshore supply vessels 30.3
- overhaul of cargo gear 17.9
- register 17.15.5; Appendix 4C
- stability 17.2.13–14
- strength 17.2.1–2
- testing 17.12; 17.2.43
- training of operators 17.2.7–12
- trucks and other mechanical handling appliances 17.10
- union purchase 17.7
- use of derricks 17.6–7
- use of lifting plant 17.2
- use of stoppers 17.8
- winches 17.3–4

Lighting 9.3
- defective 9.3.5
- on offshore supply vessels 30.1.6–7
- portable 9.3.8
- portable, for welding and flamecutting 13.1.3
- temporary 9.3.8
- transit areas 9.3.2
- work benches and machines 12.4.4; 12.4.7
- working areas 9.3.2

Liquefied gas carriers 29.3

Litter 1.4.8; 9.2.6

Loading of cargo 27.3

Logbook
- appointment of safety officials 4.2.4; 4.3.5; 4.4.5

Lung cancer 1.2.19

Machinery
- overhaul of 22
- painting in vicinity of 14.1.6
- use of permit to work system 7.1.5

Machinery spaces
- boilers 20.2
- fire prevention 2.5
- general precautions 20.1
- refrigeration machinery 20.4
- unmanned spaces 20.3

Machines 12.4
- checking before use 12.4.5
- eye and face protection 12.4.11
- guarding 12.4.2–5
- operation 12.4.1; 12.4.5
- position of controls and switches 12.4.4
- removal of swarf 12.4.9
- repairs 12.4.6
- stopping and starting 12.4.15
- V-belt 12.4.16

Machine tools 12.4

Main engines
- overhaul of 22.8

Making fast 16.3

Malaria 1.2.6

Manual handling 11
see also Lifting

Manual lowering 11.2.10

Marline spikes 22.1.15

Masters
- duties in respect of dangerous spaces 10.15.1
- duties towards safety officials 4.9
- responsibilities in respect of lifting plant 17.2.1; 17.2.4–6
- responsibility for accident reporting 4.10.3
- responsibility for safe means of access 8.1.1
- responsibility for safe movement on board 9.1.1

Matches
- in dangerous spaces 10.8.2

Materials 12

Means of access 8
- access by boat 8.9.1–2
- accommodation ladders 8.2.1–2; 8.2.4; 8.4
- construction 8.2
- freedom from obstruction 8.5.2
- gangways 8.2.1–3; 8.4
- lighting 8.5.1
- maintenance 8.3
- positioning 8.4
- unconventional 8.9

Means of Access Regulations 8.1.1

Medical First Aid Guide (MFAG) 27.2.5; 29.4.4; Appendix 2 (under 'IMO publications')

Merchant Shipping Regulations
- MS (BCH Code) Regulations 29.4.4
- MS (Closure of Openings in Hulls and Watertight Bulkheads) Regulations 3.2.8
- MS (Code of Safe Working Practices) Regulations 1.1.8
- MS (Crew Accommodation) Regulations 20.4.3
- MS (Dangerous Goods and Marine Pollutants) Regulations 6.7.1; 27.2.1
- MS (Entry into Dangerous Spaces) Regulations 10.15.1; 29.4.7
- MS (Gas Carriers) Regulations 29.3.1
- MS (Guarding of Machinery and Safety of Electrical Equipment) Regulations 12.4.2 et seq.; 20.1.1
- MS (Hatches and Lifting Plant) Regulations 17.1.1;
  Reg 4(2) 18.1.1;
  Reg 4(3) 18.1.2;
  Reg 4(4) 18.1.3;
  Reg 4(5) 17.2.7; 18.1.4-5;
  Reg 5 17.1.1;
  Reg 6 17.2;
  Reg 7 17.12;
  Reg 8 17.13;
  Reg 9 17.14;
  Reg 10 17.15
- MS (Health and Safety: General Duties) Regulation 1.1.9
- MS (IBC Code) Regulation 29.4.4
- MS (Musters and Training) Regulations 3.1.1
- MS (Means of Access) Regulations 8.1.1
- MS (Protective Clothing and Equipment) Regulations 5.1.1; 30.1.10
- MS (Provisions and Water) Regulations 25.3.11
- MS (Safe Movement on board Ship) Regulations 9.1.1
- MS (Safety Officials and Reporting of Accidents and Dangerous Occurrences) Regulations:
  part 1 (safety officials) 4.1-9;
  part 2 (accident reporting) 4.10

Mercury
- in paint sprays 14.2.4

Microwave ovens 25.3.15-16

Microwave radiation 23.1.1; 23.1.3

Mooring 16.4
- to buoys 16.4
- lifejackets and lifebuoys 16.4.1; 16.4.3

Mooring lines 30.1.12

Mooring ropes 16.3.3; 16.3.9

Movement 9

Musters 3.1
- muster list 3.1.3
- muster stations 3.1.3

New entrants
- safety training 4.6.13

Nitric oxide 10.4.15

Nitrogen dioxide 10.4.15

Noise
— machinery 20.1.3-4

Non sea-going ships
— definition 4.1.6

Notices 6
— no smoking or open flames in battery compartment 24.1.2
— operation of boilers 20.2.1
— painting work below ship's side discharges 14.3.6
— safety in unmanned machinery spaces 20.3.2
— safety signs and notices 6.2; 9.4
— solvent hazard in dry cleaning plant 26.4.7
— treatment of electric shock 22.9.2
— warning, see Warning Notices

Nylon
see Polyamide

Obstructions 9.1-2
— permanent fittings 9.2.4

Offshore supply ships 30
— anchor handling 30.4.1; 30.7
— approaching rig 30.4
— cargo handling at rig 30.4
— carriage of cargo on deck 30.2
— general guidance 30.1
— lifting, hauling and towing gear 30.3
— lighting 30.1.6-7
— personnel baskets 30.5
— transfer of cargo 30.4.5 et seq
— transfer of personnel 30.5-6

Oil
— for cooking 25.3.11
— leakages 20.1.5-6
— prevention of leakage 2.5.1
— safety lamps in presence of oil vapour 22.1.17
— spillages and leakages of oil cargo 29.2.6
— spillages in transit areas 9.2.2
— toxicity of vapours 29.2.2

Oil cargo
— toxicity 10.4.10-12

Oil carriers 29.2

Oil rigs
see Offshore supply ships

Oil tanks
— overflows 20.1.7

Oil tins
— not to be kept in boiler rooms or machinery spaces 2.5.3

Openings
— guarding of 9.5

Operators
— of lifting plant 17.2.7–11; 17.2.20; 17.2.24
— of powered mobile lifting appliances 9.7
— of ships' vehicles 9.7

Outboard
— working 15

Ovens 25.3
— microwave 25.3.15–16

Overside
— painting 14.3

Oxygen
— compressed 10.5.5
— deficiency 10.1.1–2; 10.4.1–9; 10.6.5
— depletion during gas cutting 13.1.4

Oxygen cylinders
— stowage 12.8.3

Oxygen deficiency
— breathing apparatus 5.5.9–11

Paint
— not to be kept in boiler rooms or machinery spaces 2.5.3
— fumes during welding and flamecutting 13.1.4

Painting 14
— general guidance 14.1
— painting aloft, overside and from punts 14.3
— spraying 14.2

Paints 14.1.1

Pantries 25

Passengers
— safe movement 9.2.8; 9.4.2

Pedestrians
— on dockside 8.9.5
— on ferry ramps 8.9.4

Permits to work 7
— dangerous spaces 10.7 (see also 29.4.8 (chemical tankers), 29.2.10 (oil tankers) and 20.3.1 (unmanned machinery spaces)).
— on chemical tankers 29.4.8
— on oil tankers 29.2.10
— overhaul of machinery 22.1.1
— principles of system 7.1.3
— specimen permit 7.1.5
— warning notices 7.1.4
— welding and flamecutting 13.1.1

Personnel baskets 30.5

Pesticides
— in paints 14.1.1

Petroleum
— as bulk cargo 29.2

Phenol 25.1.12

Pipelines
— colour codes 6.6

Pleasure craft
— definition 4.1.7

Plugs—see Electrical plugs, Earplugs etc.

Pneumatic drills
— vibrations from 12.3.14

Pneumatic equipment 21
— repairs and maintenance 22.12

Pneumatic tools 12.3

Polyamide ropes 15.4.2 (table); 15.4.6
— overworking of 16.2.5
— splicing of 16.2.10(a)
— use of in mooring 16.2.2

Polyester ropes 15.4.2 (table)
— splicing of 16.2.10(a)
— use of in mooring 16.2.2

Polyethylene ropes
— splicing of 16.2.10(c)

Polypropylene ropes 15.4.2 (table); 15.4.5
— effect of heat on 16.2.9
— splicing of 16.2.10(b)

Portable ladders 8.6; 9.6.6–7; 15.5

Power tools 12.3
— cables 12.3.4–7
— guards 12.3.11; 12.4.2–3; 12.4.5
— protective clothing 12.3.13
— safe voltage 12.3.2
— switching off and disconnecting 12.3.12
— "white finger" 12.3.14

Pressure gauges 21.1.2

Proof loading 17.12.3–4

Protective clothing and equipment 5
— caps and helmets 5.2
— defective equipment 5.1.4
— earplugs and earmuffs 5.3
— face masks 5.4
— footwear 5.6.2–3

Protective clothing and equipment—*continued*
— gloves 5.6.1
— goggles 5.4
— handling of dangerous substances 27.2.8
— inspection of equipment 5.1.5
— lifebuoys 5.9
— limitations of protection 5.1.3; 5.1.6
— mooring 16.4.1; 16.4.3
— outerwear 5.8
— respirators 5.5
— safety harnesses 5.7.1
— training in use of equipment 5.1.6
— welding and flamecutting 13.2

Protective Clothing and Equipment Regulations 5.1.1; 30.1.10

Pumping operations
— suspension during entry into dangerous space 10.5.6

Punts
— painting from 14.3.2–3

Quayside
— safe access 8.9.5

Radar 23
— work near scanner 15.1.8

Radiation 23.1.1–4
— microwave 23.1.1; 23.1.3
— X-ray 23.1.4

Radio equipment 23
— aerials 15.1.7; 23.1.7–9
— electrical hazards 23.2
— extension runners 23.4
— valves and semi conductors 23.3

Rags
— fire prevention 2.4; 2.5.2

Ramps
— angle of vehicle ramps 9.7.7
— ferry type 8.9.4
— operators of 17.2.7–8

Rats 1.2.10

Reciprocating tools 12.3.14

Refrigerants 22.10.3–5
— leakages 20.4.5

Refrigerated rooms
— for provisions 25.6

Refrigeration machinery 20.4
— overhaul 22.10

Rescue
— from dangerous space 10.2.3; 10.11

Rescue boats 3.3.10

Rescue harness 10.9.5; 10.10.6

Respirators 5.5.3–9
— positive pressure 5.5.5
— use when handling dangerous substances 5.5; 27.2.8
— use when paint spraying 14.2.5
(see also Breathing apparatus)

Resuscitators 5.5.2

Rigs 30
— approaching rig 30.4
— cargo handling at rig 30.4
— rig anchors 30.7
— transfer from ship to rig 30.5–6

Risk assessment 1.1.11–12
— dangerous space 10.2.1; 10.3.1–7
— hazardous substances 1.5
— injury from lifting and carrying 11.1.2–3
— permits to work 7.1.3(a)
— record of assessment 1.5.7
— review of assessment 1.5.7

Rope ladders
— used for access 8.2.6; 8.6

Ropes 15.4; 16.2
— backlash hazard 16.2.8
— corrosion etc 15.4.7; 15.4.9; 16.2.4–5; 16.2.9
— hazard of standing in a bight 16.3.11
— inspection of ropes 15.4.7
— man-made rope fibres 16.2
— protection from edges of loads 27.3.14
— rope fibres 15.4.2; 15.4.4 et seq.; 16.2
— splicing 15.4.11–12; 16.2.10
— storage 15.4.3
— stretching 15.4.6
— use when aloft or outboard 15.4
— wire 16.3.4 et seq; 17.3.35; 17.3.39

Rope slings 17.12.1; 17.12.5

Rubbish
— fire risk 2.4.1
— in engine room bilges 20.1.9
— in galley 25.1.13

Rust removers 14.1.3

Safe movement 9
— general advice 9.10
— guarding of openings 9.5
— ladders 9.6
— passengers and dock workers 9.2.8; 9.4.2

Safe movement—*continued*
- transit areas 9.2
- vehicles 9.7
- watertight doors 9.9

Safe working loads (SWL) 17.2.43–47; 17.14.1–2; 17.14.5; 27.3.8
—marking of SWL 17.14.1–2; 17.14.5; 22.3.1

Safety
- complaints and suggestions 4.6.2; 4.9.3
- co-operation 4.1.2
- duties of employer, master and crew 4.1.1
- general duties 1.1.9
- information 4.9.5
- inspections 4.6.3; 4.6.14 et seq.
- policy 4.6.13
- recommendations 4.6.4

Safety committees 4.4; 4.8
- agenda 4.4.9–13
- chairman 4.4.5; 4.9.4
- composition 4.4.2; 4.4.4; 4.4.7
- frequency of meetings 4.4.8
- minutes 4.4.14–15
- secretary 4.4.6

Safety harnesses 5.7; 15.1.3; 15.5.10
- on offshore supply ships 30.1.10
- use in cargo spaces 19.1.2

Safety helmets 5.2
- on offshore supply ships 30.1.11
- use when overhauling engine room machinery 22.2.1

Safety lamps
- use in presence of oil vapour 22.1.17

Safety nets 8.3.1; 8.7; 15.1.3
- use in cargo spaces 19.4.4

Safety officers 4.6
- accident investigation 4.10
- advisory role 4.6.22; 4.6.27
- appointment 4.1.4; 4.2
- duties 4.1.4; 4.6
- knowledge of safety legislation 4.6.27
- power to stop unsafe work 4.6.26
- relationship with colleagues 4.6.22–25
- termination of appointment 4.5.1
- training 4.9.2

Safety officials 4

Safety representatives 4.7
- election 4.3
- eligibility 4.3.1
- number 4.3.4
- powers 4.7
- relationship with colleagues 4.7.3
- termination of appointment 4.5.2

Safety signs 9.4

Salt tablets 1.2.14

Sanding
— protective clothing 12.4.12–13

Saws 12.2.8
— for catering 25.5

Scalds
— hazard in laundries 26.2

Semi-conductors 23.3

Shackles
see Lifting gear

Signallers 17.2.24 et seq.; 27.3.4–6

Signs 6
— information 6.2.5
— precaution 6.2.3
— prohibition 6.2.2
— supplementary 6.2.6
— warning 6.2.4

Skin diseases 1.2.7–1.2.13

Slings
— use when lifting cargo 27.3.8 et seq.

Slippery surfaces 9.2.2
— in machinery spaces 22.2.5

Slips and falls 5.7
— hazard in galleys 25.2

Smoke
— escape from 3.4.7

Smoking
— and asbestos 1.2.19
— and fire prevention 2.1
— and fumigation sprays 1.4.13
— effect on paint vapours 14.1.5
— prohibited in battery compartments 24.1.2
— prohibited when handling explosives and flammables 27.2.7
— rules when carrying bulk oil cargoes 29.2.5
— warning notices 2.1.2

Sodium hypochorite 25.1.12

Solvents
— for cleaning electrical equipment 22.9.5
— for cleaning machinery 20.1.11
— for degreasing radio equipment 23.1.5
— from tank coatings 10.4.18
— in dry cleaning 26.4

Solvents in paints 14.1.1

"SORADO" Regulations 4
— application 4.1.5
— exemptions 4.1.8
— Part 1 (safety officials) 4.1–9
— Part 2 (accident reporting) 4.10

Spirit lamps 12.6

Spontaneous combustion 2.4

Spray painting 14.2
— equipment 14.2.1–2
— hose pressure 14.2.9
— protective clothing 14.2.3
— spray nozzles 14.2.6–7

Sprinkler system
— interruption of water supply 22.1.2

Staging 15.2

Stanchions 8.4.4

Steam
— hazard in laundries 26.2.1

Steam boilers 25.3

Steam pipes 20.1.2

Steering gear
— overhaul 22.11

Stoppers 17.8

Storage batteries 24

Stoves 25.3

Stowage
— of gas cylinders 12.8.3
— of working equipment 12.4.8

Substances hazardous to health 1.5
— dangerous goods 27.2

Sulphur dioxide 10.4.15

Sulphuric acid 24.2

Sunshine
— exposure 1.2.15–16

Survival craft
— drills 3.3
— procedures for lowering 3.3.13–14
— radio equipment 3.3.7

Swarf
— in machine areas 12.4.9

Tank coatings
— solvent vapours 10.4.18

Tankers 29
see also Bulk liquid cargoes

Tarpaulins
— use with hatch covering 18.4.1
— dangerous on partly opened unguarded hatches 19.3.3

Testing
— of lifting plant 17.12

Tools 12
— compressed air 12.7
— electrical, pneumatic and hydraulic 12.3
— fixed 12.4
— hand 12.2
— safe use when aloft 15.1.11-13

Towing 16.6
— tow ropes 16.6.3; 16.6.7

Toxicity
— of hydrocarbons 10.4.10-12
— of inert gas 10.4.15

Training
— entry into dangerous spaces 10.14
— lifting plant operators 17.2.7-12
— new entrants 4.6.13
— safety officers 4.9.2
— use of life-saving appliances 3.1.4
— use of protective equipment 5.1.6

Transit areas 9.2

Trucks 17.2.47; 17.10
— marking of safe working load 17.14.1

Tyres 17.2.21

Union purchase 17.7

Vaccination 1.2.5

Valves
— electronic 23.3

Vehicle ramps 8.9.4

Vehicles
— drivers of ships' 9.7.1
— traffic control 9.7.5

Ventilation
- dangerous spaces 10.5.4; 10.9.1
- during painting and paint drying 14.1.4
- of battery compartments 24.1.1
- of cargo spaces 19.1.1
- of drying cabinets 2.2.9
- of laundries 26.2.2
- of refrigerated compartments 20.4.3-4

Vibration
- from tools 12.3.14
- "white finger" 12.3.14

Visual alarms
- in high noise level areas 20.1.4

Visual display units (VDUs) 23.5

Walkie-talkies 16.1.3

Walkways 9.2.1
- equipment stowed on 9.2.5

Warning notices
- no smoking 2.1.2
- slippery access area 8.5.2
- work near propeller aperture 14.3.5
- work near radar scanner 15.1.8
- work near radio aerials 15.1.7
- work near ship's whistle 15.1.5
- work overhead 15.1.11

Washing
- before food handling 25.1.2

Washing machines 26.3

Washrooms
- drainage 9.8.1

Waste materials
- disposal 25.1.13
- fire risk 2.4.1

Watertight doors 9.9
- use of notices 9.9.3

Welding 13
- in dangerous spaces 10.8.2
- electric 13.4-5
- fire and explosion precautions 13.3
- fumes 13.1.4
- gas welding 13.6
- general guidance 13.1
- inspection and repair of equipment 13.1.5
- protective clothing 13.2
- screens to protect open hatches 13.3.3
- supervision 13.3.7
- testing for gases 13.3.6
- ventilation 13.1.4
- welding helmets 13.2.2

Winches 17.3-4
- safe use 17.3-4
- to lift beams, pontoons and slab hatches 18.4.6
- use to move heavy loads 17.2.31
- use when making fast 16.3.8; 16.3.12; 16.3.18

Wind
- hazard during lifting operations 17.2.40

Wire ropes
- inspection 17.2.39
- lubrication 16.3.4; 17.2.39
- splicing 17.2.35

Wirebrushing
- face protection 12.4.11

Wiring 6.4

Wood chisels 12.2.6

Work aloft 15
- general guidance 15.1
- cradles and stages 15.2
- portable ladders 15.5
- ropes 15.4

Work outboard 15
see also Work aloft

Working clothes 1.3; 5.1.1
- painting near machinery 14.1.6

Workshop 12
- machines 12.4
- working areas 12.4.8

X-ray radiation 23.1.4